MUSHROOMS FOR LIVELIHOOD

PROF. (MRS.) VIJAYA KHADER, PH.D

Post Doctoral Fellow (CSIR Fellow)
Fellow of Association of Food Scientist and Technologist (AFST(I))
Former Dean, Faculty of Home Science (ANGRAU)
Board of Governor Farm and Rural of Science Foundation (FRSF)
Member of Third World Organization for Women in Science (TWOWS)
Organization for Women in Science for the Developing World (OWSDW)
Agriculture & Agro Industry Committee Member, Federation of Andhra Pradesh Chambers of Commerce & Industry (FAPCCI)
Steering Committee Member, Millet Net Work of India (MINI)

Former - Prof. & Principal
College of Home Science
Andhra Pradesh Agricultural University
Bapatla, Guntur District
Andhra Pradesh

KALYANI PUBLISHERS

LUDHIANA - NEW DELHI - NOIDA (U.P.) - HYDERABAD
CHENNAI - KOLKATA - CUTTACK - GUWAHATI - KOCHI - BANGALORE

KALYANI PUBLISHERS

Head Office :
B-I/1292, Rajinder Nagar, **Ludhiana**-141 008
e-mail : kalyanibooks@yahoo.co.in

Administrative Office :
4779, Ansari Road,
23, Daryaganj, **New Delhi**-110 002

Works :
B-16, Sector 8, **Noida** (U.P.)

Branches :
1, Mahalakshmi Street,
T. Nagar, **Chennai**-600 017

110/111, Bharatia Towers Badambadi
Cuttack-753 009 (Odisha)

3-5-1108, Narayanaguda, **Hyderabad**-500 029 (A.P.)

10/2B, Ramanath Mazumdar Street,
Kolkata-700 009

1st Floor, Saraswati Road, Pan Bazar,
Guwahati-781 001
Convent Road, **Kochi**-682 035
No. 24 & 25, Ist Floor, Hameed Shah Complex,
Banpet Main Road, **Bangalore**-560 002

Second Revised and Enlarged Editon 2013

Reprinted, 2018

Typesetting & Diagrams by:
(MUSHROOM)
Datalink Computer
213-Sangam Complex, Milap Chowk,
Jalandhar-144 001.

ISBN 978-93-272-2798-7

PRINTED IN INDIA
At Kalyani Printings B-15, Sector 8, NOIDA (U.P.)
and Published by Mrs. Usha Raj Kumar for
Kalyani Publishers, New Delhi-110 002

PREFACE

Increasing food production to meet our country's needs is essential. With the existing conventional agricultural products and methods we may not be able to cope up with the food problem. There is an urgent need to identify, cultivate and popularise new foods which cannot only augment the food deficits but also are nutritious and easily available.

Mushroom along with yeast and algae are often referred to as alternative sources of food. However, among these mushroom cultivation is unique in the sense that it is the most efficient and economically viable biotechnology process for the conversion of lignocellulose waste materials into high quality protein food and this will naturally open up new job opportunities especially in rural areas.

Mushroom growing is becoming very popular because it not only meets the dietary requirements but also adds to the income of especially those growers having no or insufficient land. It is considered to be a very rewarding and fascinating hobby particularly for the retired persons and also house wives who can grow mushrooms in small boxes or other containers in their houses while attending to household chores.

Of the many mushrooms that can be cultivated, only three kinds namely button mushroom (Agaricus bisporus) Paddy Straw Mushroom (Volvariella Volvacea) and Oyster Mushroom (Pleurotus sajorcaju) are suitable for growing in India where suitable environmental conditions exist.

There are many popular publications on the subject but either they are not comprehensive to include the well-known edible mushrooms. They deal with the sophisticated expensive technology of the west, which has little relevance to India.

There has been a spontaneous demand from the public, by the literature which would help them to grow mushrooms. I hope this literature provides to the prospective growers, guidance for mushroom culture on cottage scale under natural environmental conditions.

The opening chapters of the book entitled, "Mushrooms for Livelihood" provide an overall view of preparation of pure culture and preparation of spawn, mushrooms cultivation in India and types suitable for cultivation. This is followed by cultivation of four varieties of mushrooms along with the economics and pests and diseases of mushrooms. The final last chapter deals with preservation of mushroom and mushroom recipes. Some of the publications are also enclosed for the benifit of the readers. I hope this book will be found useful by the mushroom growers and those who are interested in mushrooms in this country.

2013 **Prof. (Mrs) VIJAYAKHADER, PHD**

CONTENTS

CHAPTER 1

STATUS OF CULTIVATION OF MUSHROOMS IN WORLD AND INDIA

CONTENTS

Ref: *P.K. Khanna and L.S. Kapoor (2007) Manual on mushroom production. Dept. of Microbilogy college of basic sciences and Humanities, PAU, Ludhiana PP. 7-10.*

Status of mushroom cultivation

Since early times, mushrooms have been treated as a specialty item of food. The Greeks regarded mushrooms as providing strength for warriors. The Pharaohs prized them as a delicacy and the Romans called them the "Foods for the Gods". The Chinese termed mushrooms as "Elixir of life". In the beginning these mushrooms were collected from their wild habitats, but with the passage of time there were attempts to cultivate them. The earliest cultivated mushroom was a specie of *Auricularia,* about a 1000 years ago in China. This was followed by *Lentinus edodes* (900 years ago). *Agaricus bisporus* the most popular variety, called the button mushroom was first cultivated in France in 1650. *Volvariella volvacea* commonly called the paddy straw mushroom was introduced about 300 years back in China. With the use of the latest technical advances, the production of mushrooms has increased over the past few decades and at present the total world production is about 10 million MT.

Present status of Mushroom Cultivation

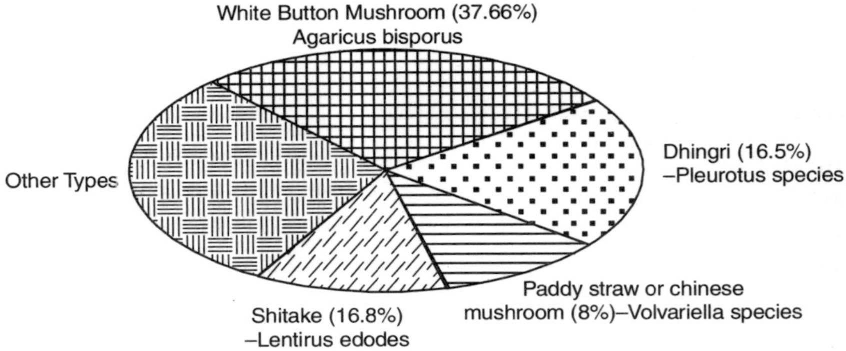

The most important cultivated varieties include White button mushroom (*Agaricus bisporus*), Shiitake (*Lentinus edodes*), Oyster mushroom (*Pleurotus* spp.), Jew ear's mushroom (*Auricularia* spp) and Chinese mushroom (*Volvariella* spp). The major producers are China, USA, Netherlands, France and Japan while the major consumers nations are Germany, Italy, France, U.K. and the United States of America.

World Production of Cultivated Edible Mushrooms

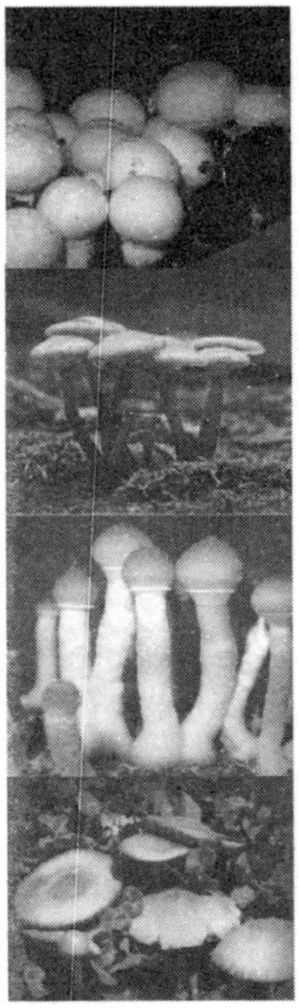

Species	% Share
Agaricus bisporus/bitorquis	37.6
Lentinus edodes	16.8
Pleurotus spp.	16.3
Volvariella spp.	8.5
Auricularia spp.	6.1
Flammulina velutipes	4.7
Tremella fuciformis	3.2
Hypsizygus marmoreus	1.1
Pholiota nameko	0.6
Grifola frondosa	0.3
Others	4.8
Total	100.0

Mushroom cultivation in India

Mushroom cultivation in India is about 4 decades old with a rapid increase in production over the last 10-12 years. From a meagre production of about 4,000 tonnes in 1985, India now produces more than 80,000 tonnes of mushrooms annually of which about 90% is of the button mushroom. The remaining 10% production is accounted for by oyster mushroom and Chinese mushroom. The major chunk of national mushroom production, primarily button mushroom, comes from a number of industrial units located in different states.

Mushroom cultivation in Punjab

Four mushroom varieties are recommended by the Punjab Agricultural University, Ludhiana for seasonal cultivation under Punjab conditions. These are *Agaricus bisporus* (button mushrom) & *Pleurotus* spp. (oyster mushroom or *dhingri*) during winter months from October to March and *Volvariella* spp. (Paddy straw mushroom) & *Calocybe indica* (milky mushroom) during summer months from April to October. Punjab is the leading producer state and contributes about 50% of the total mushroom production in the country. The state has about 400 seasonal growers along with 4 controlled-growing industrial units including M/s Agro Dutch Foods Ltd, the largest mushroom producer and exporter from India.

Type of Grower	Approx. no. of growers	Straw processed (q)	Mushroom production (MT)
Small Domestic	200	<50	
Medium Cottage	125	<100	5,000
Commerical Sesonal	70	>100	
Industrial	4	>500	35,000

Incidentally ail the mushroom growers of the state are engaged in the commercial cultivation of button mushroom, *Agaricus hisporus* only, a few seasonal growers cultivate oyster and Chinese mushrooms but at kitchen-gardening or cottage scale. Perhaps this is the reason that the later two varieties have found little favour with the consumers due to their ignorance about these mushroom types. Milky mushroom has recently been introduced for commercial cultivation in the state.

CHAPTER 2

HISTORICAL ASPETS OF MUSHROOM CULTIVATION IN INDIA

Although the methods of cultivating edible mushrooms were known to people in the various countries, its cultivation remained obscure in this country till recently.

Bose (1921) was successful in culturing two agaricus on sterilized dung. Su and Seth (1940) described the procedure for spawn production and cultivation of Volvariella. Thomas et al. (1943) cultivated the Paddystraw mushroom at the College of Agriculture, Coimbatore. There after many improvements and modifications in the growing method were worked out at several research centres in India (Asthana, 1947; Rath, 1961; Rama Krishnan et al., 1968; Gupta et al., 1970; Munjal, 1973; Purkayastha et al., 1980).

The prospects of growing the white button mushroom a temperate mushroom in India, were first discussed by Padwick (1941). This generated much interest in its commercial cultivation. Consequently, the Indian Council of Agricultural Research and the Government of Himachal Pradesh jointly started a scheme for the experimental cultivation of this mushroom at Solan in 1961 and subsequently strengthened it, by providing the services of international mushroom experts. The conversion of the Mushroom Research Project, Solan, into an enlarged ICAR Co-ordinated Mushroom Research Scheme with its main centre at Solan and the 3 subcentres at Ludhiana, New Delhi, and Bangalore gave a further fillup to mushroom research in India. On termination, the scheme became the nucleus of the National Centre of Mushroom Research and Training at Solan, the present-day premier Institute of mushroom research in India.

Some agricultural universities such as Punjab Agricultural University, Ludhiana, Chandra Sekhar Azad University of Agricultural and Technology, Kanpur, Govind Ballabh Pant University of Agriculture and Technology, Pantnagar; Tamil Nadu Agricultural University, Coimbatore and Action Research Institute, Calcutta are conducting research on mushroom cultivation. This has led to an awakening among scientists in other institutions as well and the Council of Scientific and Industrial Research has started a small unit at Srinagar.

The other mushroom which has succeeded and shown its commercial potential is pleurotus sajor-caju. The technology for its cultivation was worked out at the Indian Agricultural Research Institute, New Delhi (Jandaik and Kapoor, 1976). As compared with other edible species of pleurotus, this species exhibits a wider adaptability. Because of its fine characteristics, the mushroom is now being cultivated on a commercial scale in many countries.

Commercial production began in the United States in the 1880s. Agaricus is the leading mushroom crop worldwide and accounted for 99 percent of the 1997 United States' mushroom production. Oyster mushrooms (Pleurotus spp.) were more recently domesticated, and now rank second in world production. Shiitake mushrooms, which are very popular in Asian cultures, rank third. Many other edible mushrooms, such as, straw and wood ear mushrooms, are gaining in popularity.

Roughly 300 mushroom species are edible, but only 30 have been domesticated and 10 are grown commercially. Button, oyster, and shiitake mushrooms make up about 70 percent of the world's production. During the past 30 years, mushroom production worldwide increased twenty-fold, with much of that increase occurring in the 1980s and 1990s. Increased demand for specialty mushrooms (everything besides Agaricus) has been particularly strong. Asian countries continue to dominate world production and consumption

Management

Commercial mushroom production requires high levels of management input and skill. A common mistake new grower's make is to believe that growing mushrooms is easy. Each species requires specialized treatment to produce consistent yields of high-quality, marketable mushrooms. Another common mistake is to start too large and diversify too soon. As mentioned earlier, mushroom growing is labor intensive. It is easy to quickly become overwhelmed with the physical requirements of mixing and sterilizing substrates, ordering and inoculating with spawn, maintaining environmental controls, harvesting and processing mushrooms, marketing, business management, and many other tasks that go with a commercial enterprise. Trying to learn a single crop is difficult enough, and mastering several different mushroom crops at once may be impossible. Some spawn suppliers offer starter kits and instructions. Using small starter kits will allow you to gain some experience with different mushroom crops with minimum investments in time and money. Keep detailed production and financial records to evaluate which crops show commercial promise. As with any other crop species, not all mushrooms are created equal. Different strains or lines of shiitake, for example, vary in color, size, shape, firmness, cultural requirements, and yields. Only the largest mushroom growers produce their own spawn. Spawn culture is highly technical and requires specialized facilities and equipment. Most growers rely on companies that specialize in producing high-quality spawn for their culture material. If you are not already experienced in mushroom production, start small and expand slowly. Take time to learn all you can about growing and selling mushrooms while you gain some practical experience. Study the market and decide which types of mushroom crops and production systems would be enjoyable, feasible, and profitable for you. Join growers' organizations and subscribe to newsletters about mushrooms. Universities sponsor conferences and workshops on specialty farming in general and mushroom farming in particular.

Marketing

The greatest challenge all specialty farmers face is marketing. Deciding what to grow, where and how to grow it, who makes up the target market, and how to package and advertise are just a few of the things that go into marketing? A thorough study of mushroom production and marketing is imperative before buying equipment and starting even a small-scale operation. The demand for specialty mushrooms is huge, particularly in Asian countries. Trying to market internationally, however, is beyond the resources of most small and medium-sized companies. China produces nearly 1.5 billion pounds and Japan more than 300 million pounds of shiitake each year. Likewise, national markets in the United states are dominated by large companies and produce brokers. Most small-scale mushroom farmers in the United States focus on local markets. Specialty mushrooms are best known and most widely used among certain Asian cultures, and local sales may be best in areas with large populations of Asian-Americans. Mushrooms are sold fresh, dried, or processed. According to the U.S. Department of Agriculture, most specialty mushrooms grown in the United States are sold fresh. For wholesale, consider locally-owned or operated groceries, restaurants, and health food stores. You can make direct sales to customers through farmers markets, subscriptions, and on farm sales. Forming a cooperative with other growers in your area can improve marketing by increasing quantities and variety. Before deciding on any market strategy, thoroughly explore local, state, and federal regulations that will affect your growing, processing, and shipping.

In conclusion

Specialty mushrooms offer small-scale growers opportunities, however, there are risks. Do your homework before investing in land or production facilities. Read several books by different authors to get a balanced view of mushroom farming. Treat claims of quick and easy profits with great skepticism. Visit several mushroom growers outside your area to gain a grower's-eye-view of what it is like to produce and market mushrooms. Join mushroom growers' organizations and attend conferences and workshops. Calculate your costs for starting and operating a mushroom farm, including labor costs. Determine break-even points and the time you will need to recapture your investment. Be conservative in estimating yields, sales, and profits. Plan on supporting your mushroom operation with savings or off-farm work until you pass the break-even point. In short, give yourself every chance for success.

CHAPTER 3

PROGRESS OF WORK ON MUSHROOM IN COLLEGE OF HOME SCIENCE, (APAU) BAPTALA

Dr. (Mrs.) Vijayakhader, started Mushroom Cultivation as hobby with Velveriella Valvacea (Paddy Straw Mushroom) at Sardar Patel University, Gujarat during 1977-78. After joining APAU with an idea of helping rural women the cultivation of Pleurotus (Oyster Mushroom) was initiated in the month of April, 1987 in Bapatla in two voluntary organisations i.e. CASA (Hanumantha Nagar) and GRAMASIRI (Nandiraju Thota). One training programme was organised from 29th April to 1st May, 1987 in collaboration with CASA to thirty women social workers without any financial commitment from APAU. The idea of Mushroom cultivation was well accepted by public and attracted the media. Number of Radio Programmes were organised. Handouts in Telugu and English were prepared and distributed free of cost. Brochures inTelugu were printed under N.S.S. funds and distributed freely. APAU was kind enough to print a technical bulletin entitled, "Money from Mushroom" during 1987. With the increasing awareness in the field to fulfill the need of the rural families one book on "Mushroom Cultivation" was written in Telugu and the same was released by the then Minister for Agriculture Sri Nedurumalli Janardhana Reddy in the month of March, 1990. Fifteen rural families have got bank loans to cultivate mushroom (Rs. 5,000 each) at Hanumantha Nagar (adopted village of CASA).

One T.V. Programme was shooted at Hanumantha Nagar on 4.11.87 and telecasted more than 20 times so far. Minister of Human Resources Development, Government of India has financed the workshop on "Transfer of Technology for Various Programmes on Socio Economic Development for Rural Women" from 5th to 9th July, 1990 at

College of Home Science, Bapatla where technology on Mushroom Cultivation and preservation is one of the major aspect. The authorities of APAU have given financial support (Rs. 15,000) to organise number of training programmes to the interested people from different districts of Andhra Pradesh.

So far 350 persors from Guntur, Krishna, Visakhapatnam, Nellore, West Godavari, Hyderabad, Kurnool and East Godavari districts were trained. Nearly 35 members have established Mushroom Production Units on their owm.

One book entitled, "Puttagodugulu" in Telugu was printed in APAU Press with Colour Photographs.

District Collector was kind enough to provide Rs. 27,000 under IRDP funds to establish the Spawn Production Unit. Accordingly one Autoclave, one Ultra-violet Lamp, Glass Ware, Gross Cutter etc., were purchased. Recently one Video Cassette was prepared on the simple technology of cultivation of all the three varieties of mushrooms in order to help the growers.

Audio-visual Research Centre has shooted one video programme on pure culture preparation and cultivation of oyster mushroom at Vijaywada Krishna District and Naidugudem West Godavari District. The programme was telecasted on August 5th and July 29th under UGC programme in the year 1992.

CHAPTER 4

PROSPECTS OF MUSHROOM CULTIVATION IN INDIA

The Mushroom industry is gradually taking root in India but the pace is rather slow because of insufficient scientific support and inadequate training programmes. The popularity of mushrooms unit very high in Andhra Pradesh during the last few years.

At present, there are no organized spawn producing units and, therefore, spawn supply to growers is limited and uncertain also there is no advisory service. A few research workers engaged in the activity have to attend to research, training and advisory service, with the result that they are unable to do full justice.

Mushroom cultivation has a special relevance to India because straw and plant residues are abundantly available with our farmers and our environment is conductive to the cultivation of mushrooms. Hence the future of this industry is bright. The awareness for the mushrooms among the general consuming public has been created. Swaminathan (1982) has advocated the adoption of the small farm approach in developing agriculture production. This approach has been adopted for mushroom growing by China, Japan, South Korea and Taiwan with significant success. Mushroom growing as a cottage industry is quite valid for Indian growers also because majority of them cannot afford the expensive mushroom houses equipped with boiler and environment— controlling gadgets. From the experience, till date, it has been seen that cultivation of Pleurotus Spp. has proved to be the most popular among the people because of its comparative case in cultivation. Its low capital investment and high yields that obtained even under uncontrolled rural conditions. Another factor which has led to its increased popularity is the fact that it lends itself easily to dehydration and hence can be stocked for longer periods of times as compared to Agaricus Spp. or Volvariella Spp.

The Government of Andhra Pradesh has been emphasising the importance of Agro-based exports due to the advantages of rural employment generation and gainful utilization of natural and farm resources. Mushroom cultivation is emerging in India, as a promising Agro-based, laid-independent enterprise due to flexibility in its operation.

The commercial cultivation of mushroom in Andhra Pradesh is of a recent origin and is restricted to three varieties namely, white button mushrooms (Agarious bisporus), Paddy Straw Mushrooms (Volvariella Spp.) and Oyster Mushroom (Pleurotus Spp.). Bulk of India's production is from Himachal Pradesh and Jammu and Kashmir. It is also gaining popularity in Nainital, Almora, Mussoori, Nilgiris, Coimbatore, Bangalore, Mysore and New Delhi.

As per the estimation of Prakash (1986) by using one tonne of paddy straw 100 kg. of Oyster Mushrooms could be produced around Bangalore, at a cost of around Rs. 700. The cost of paddy straw or 35% of the expenditure followed by wages (21%) and seed material (17%). At an average price of Rs. 15/- per kg. a net income of Rs. 800 was realised with an approximate benefit cost ratio 1.90. Small scale production of mushrooms seem to have remunerative benefit cost ratios making it cost suitable for small producers who can provide gainful employment to his family labour.

Market trends indicate the consumer preference towards white button mushrooms which realize higher price than other varieties. For a successful and profitable cultivation of white button mushroom proper environmental conditions (temperatures, relative humidity and aeration) and good hygiene are very vital. Hence, non-recurring expenditure on cropping rooms and in conditioning will be higher and from about 50 percent of the installations cost. Under recurring expenditure electricity, coal, firewood and water contribute nearly 40% of the expenses. The next item under recurring expenditure is the cost of labour (23%).

The non-recurring expenditure in Oyster Mushroom Cultivation is comparatively lesser than the button mushroom, as it does not require much protection. Under recurring expenditure labour (26%), paddy straw (22%) and seed material (16%) are the major items in the case of Oyster Mushroom. Paddy straw mushroom can be easily grown in thatched huts or brick store room. Hence, the installation cost is nil. Paddy straw is the major item of input and accounts for more than half of the expenditure.

At present there is an appreciable demand for the white button mushrooms. This mushroom requires a higher initial investment and quality control compared with the other two varieties. It may be advisable to produce this variety for earning foreign exchange. In order to provide self-employment opportunity to rural people, the production of Oyster and Paddy Straw Mushroom may be suitable.

Mushroom cultivation is labour intensive, about 2250 mandays of employment are generated in production of 9450 kg. of white button mushroom. About 1000 man days are required to produce 4800 kg. of oyster mushroom and 50 mandays of labour are required to produce 360 kg. of paddy straw mushrooms (Indian Institute of Horticultural Agricultural Research, 1986).

India has all the requisites of low cost labour, favourable temperature, plentiful supply of straw, spawn and other inputs to become a leading mushroom producer. The Agricultural and Processed Food Products Export Development Authority of the Ministry of Commerce started thinking in the direction of development of an export trade in cultivated mushrooms.

India needs to learn from the experience of countries like Taiwan in Mushrooms production. Taiwan encouraged production of mushrooms in small units in the rural sector, offering export incentive, technological know how and a good market net work to top the global export potential, as a result, in a brief period of six years Taiwan's production went high around mid sixties that it become the leading producer of mushroom in the world has also maintained it.

CHAPTER 5

INTRODUCTION

CONTENTS

Mushroom is a fungus but every fungus is not a mushroom. Large macroscopic fleshy fungi are known as mushroom. They are very unlike green plants because they lack chlorophyll and therefore depend on the performed food for their nutrition.

The mushroom that lives on dead matter is called a saprophyte. Some mushrooms may be parasites, drawing their nutrition foom living matter, and some facultative able to live as saprophytes as well.

Mushroom growers in India are, however, concerned with only three specific genera of the sub-division of Basidiomycotina for commercial exploitation of the technology of mushroom cultivation. These genera include Agaricus, Pleurotus and Volvariella. Agaricus is commonly known as button mushroom, Pleurotus as Oyster mushroom and Volvariella as paddy-straw mushroom. The cultivation of mushrooms in India has earned some considerable popularity in recent years with reference to the two Agaricus species *(Agaricus bisporus and Agaricus bitorquis), the two pleurotus species (Pleurotus sajor-caju and Pleoro-tus citrinopileatus) and the two Volvariella species (Volvariella diplasia and Volvariella Volvacea).*

As stated earlier mushroom is the term applied to the fleshy fruiting bodies of fungi. Under natural conditions, the mycelium lies buried in the soil or in the substratum. When conditions are favourable, it forms the reproductive structure generally called the mushroom. This is only a stage in the life cycle of the fungus. They differ in form and size. They are the part of the fungus which produces the spores, which on dispersal serve to reproduce the species in other suitable habitats. When a spore settles on a suitable substrate if atmospheric conditions are in harmony, it germinates. The spore produces a very fine filament of germ tube, which grows outward in search of nutrients, soon it branches in all directions, eventually forming a complex cobweb like system. This is the vegetative phase of the organism, the mycelium, from which the fruit body develops.

Mushrooms are good supplement for protein lacking diets and can be easily cultivated indoors and marketed profitably. They are a fair source of vitamin C, niacin pantothenic acid, folic acid, calcium, phos-phorus and iron. Being low in carbohydrates and fat, mushrooms can be a good for diabeties and obese people.

Popularly called "Vegetable Meat", mushrooms have twice the quantity of protein in leafy vegetables. Mushroom is an edible fungus. It does not have chlorophyll and hence cannot produce its own food. Instead, it obtains its nutrition from the medium on which it grows. The medium may be living or dead plant material like agricultural waste.

The edible forms of mushrooms have flushy bodies and comprise a cap and stalk. The portion which remains under the surface of the growing medium—the stalk-resembles roots of higher plant and grows extensively even before the edible portion is formed above the surface. The stalk consists of cotton like threads called mycelium which penetrate partly the decomposed growing medium to obtain nutrients as the roots of other plants do.

As in any agricultural crop, mushrom cultivation also needs good seed material. This seed material is called 'spawn'. Spawn has to be prepared by culturing pure fungus of the desired edible variety of mushroom. This seed material has to be preserved and used under utmost hygienic conditions. It is always desirable that the culture should be prepared in well established laboratories. However, if similar conditions can be arranged, mushroom spawn culture can be prepared individually by farmers.

Types Suitable for Cultivation In India

At present 3 varieties of Mushrooms are being cultivated in India. They are;

1. Oyster Mushrooms

The Oyster mushrooms are called as such as they look like Oysters. They can further be classified based on colour as Gray Oyster Mushrooms and the White Oyster Mushrooms (Fig. 1)

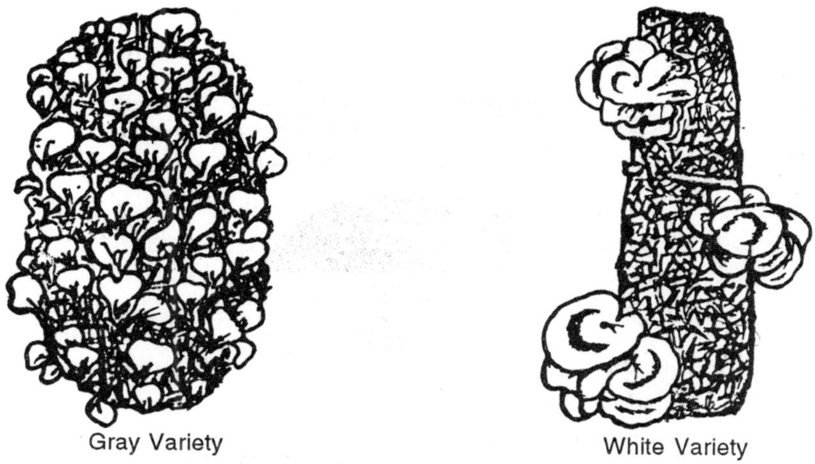

Fig. 1. Oyster Mushrooms

Oyster mushroom can grow at moderate temperatures ranging from 22° to 28°C. Therefore, it is suitable for most of the places in India. It is a familiar item in the menu of most hotels in Bangalore where it is being grown commercially.

2. Paddy Straw Mushrooms

The Paddy Straw Mushrooms are extensively grown on paddy straw, hence are known by that name (Fig. 2)

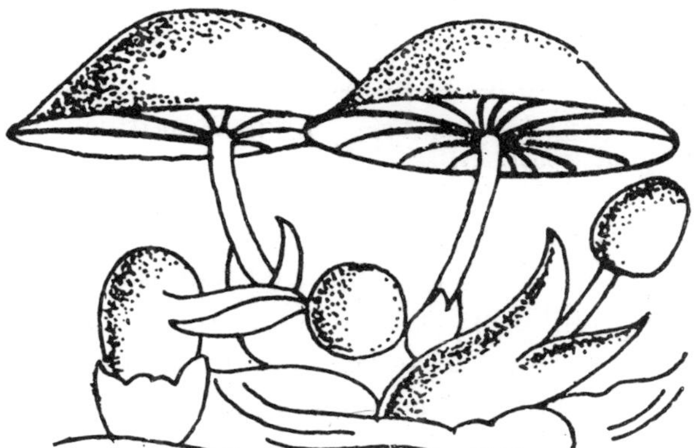

Fig. 2. Paddy Straw Mushroom

The tropical paddy-straw mushroom is suitable for growing in most parts of India. Even then it is less attractive commercially owing to very low yield per unit weight of the substrate and an extremely short self-life. But, as a kitchen-garden crop it is preferred because it is very delicious and nutritious.

Button Mushroom (Agaricus Bisporus)

The button mushroom derives its name from its shape when it is still quite young. However, the button stage is only a passing stage in the development of the mushroom which becomes umbrella-like on maturity. This mushroom has a pleasing, distinct taste and aroma. The principal parts of a mushroom are illustrated diagrammatically (Fig. 3). Market trends indicate the consumer preference towards white button mushrooms is discused in depth.

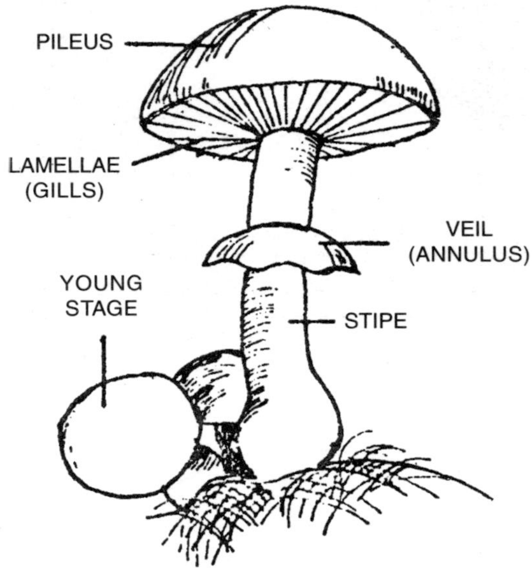

Fig. 3. Various parts of Agaricus bisporus.

Various parts of Agaricus bisporous

The fruiting body or the mushroom as it is called, consists of a stem (stipe), which supports an expanded, umbrella-shaped cap (pileus). On the underside of the pileus are the gills or lamellae, which in the young stage are enclosed by a membrane that extends from the margin of pileus to the stipe. This membrane is known as partial veil and it usually tears around the margin of the pileus as the latter expands but remains attached to the stipe where it forms a ring (annulus). Other characters can be observed only through the microscope.

A musnrom reproduces by spores produced in enormous number on the 2 faces of the gills or the lamellae on the underside of the pileus. The spores are too small to be seen with a naked eye.

The knowledge of conditions favouring, spore, germination is basic for the establishment of mushroom culture.

When a spore germinates it puts forth a germtube (a slender thread called hypha, (Fig. 4.) This thread grows rapidly, develops cross-walls or septa and becomes many celled then branches repeatedly producing a mat of hyphae termed the mycelium (Fig. 5). This, the vegetative part of the fungus, produces the furiting body or mushroom. Each cell of the mycelium has two nuclei. The vegetative mycelium may produce secondary spores which may be thin-walled or thick-walled. The latter are termed as chlamydospores.

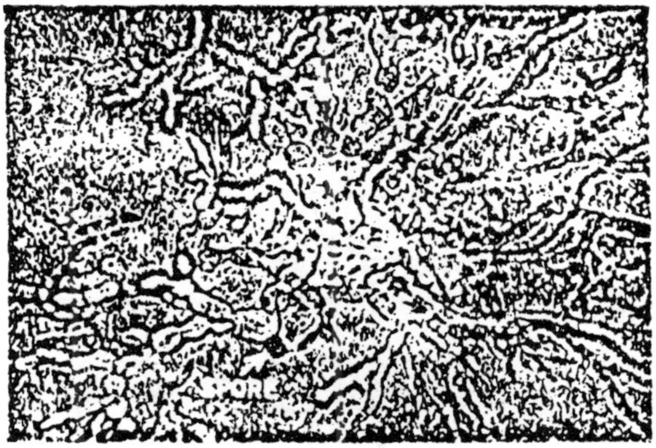

Fig. 4. A germinating spore forming mycelium.

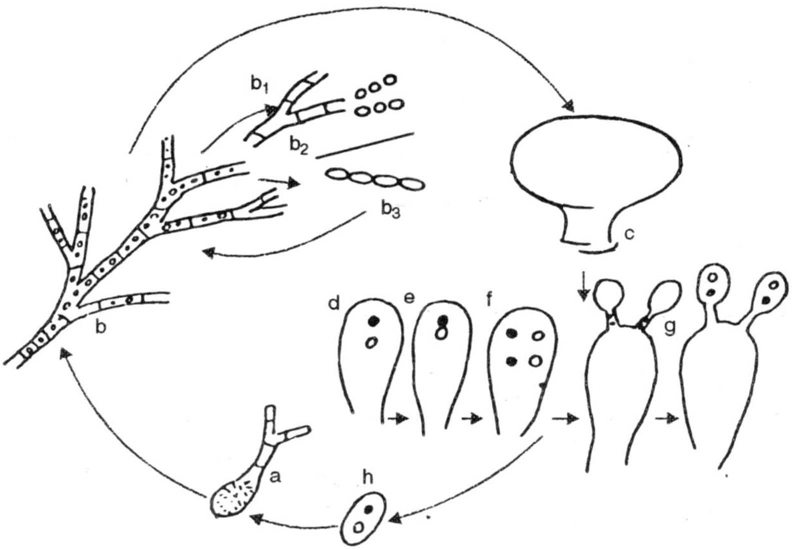

Fig. 5. Life-cycle of agaricus bisporus; a, Germtube; b, vegetative mycelium-b_1 septate hyphae; b2 secondary spores; b3 chlamydospores; c, fruiting- body initials-pins or primordia; d, young basidium; e, fusion of nuclei; f, mitosis of 4 haploid nuclei; g, a pair of nuclei passes through the sterigmata to the developing spores; h, a spore.

CHAPTER 6

PREPARATION OF CULTURE AND SPAWN

CONTENTS

Pure culture is usually obtained in 2 ways from spores or mushrooms tissues.

Eventually, the fruiting-body initials (also called pins or priomordia) are formed which develop into a mature fruit-body which forms numerous lamellae on the under-surface of the pileus. The structure of the lamellae is of special interest because it is here that the spores are produced. If a transverse section of a lamella is examined it is seen to be roughly wedge-shaped or triangular. The spores are produced on the outer surface in a definite layer called the hymenium, whereas the central part is composed of more or less interwoven hyphae called the trama. Basidia from a palisade layer which constitute a hymenium. The basidia are the cells on which the spores are produced. Each one has 2 little apical projections termed sterigmata and a spore develops on the tip of each sterigma. In the beginning the young basidium has two nuclei which fuse to form a diploid fusion nucleus. Meiosis follows and results in 4 haploid nuclei. A pair of nuclei passes through the sterigma to the developing spores. The mature spore has thus 2 nuclei. The 2 nuclei in each basidiospore may be of one sex only or of different sexes but non-compatible, or they may be of opposite sex and compatible. The mycelium derived from the spores having non-compatible nuclei remains sterile and no fruiting bodies are produced whereas in the other case, mycelium from a single spore is self-fertile (one that readily produces fruiting bodies and technically it is called secondairly homothallic). About 79% spores contain non-sister nuclei, which are compatible and the cultures derived from them are self-fertile.

Spore Culture

The spores collected aseptically or treated with a suitable sterilant such as chloroform are very carefully germinated on a nutritive medium such as malt-extract agar, which assures the maximum and most vigorous germination. The mycelium from the germinated spores is transferred to test tubes containing suitable nutritive agar medium. These are then kept in an incubator at 24° ± 1°C. Pure culture obtained in this manner is used for making spawn for production test. Pure cultures obtained from spores show a lot of variation in quality and yield because of genetical reasons. Therefore, spore cultures are not preferred for mushroom cultivation.

Tissue culture To prepare the cultures from tissues, selected mushroom of the best quality is first cleaned and dipped in a diluted disinfectant such as 5% formalin or sample wiped with rectified spirit and quickly passed through the flame. Under sterile conditions, pieces of the interior tissues of pileus or stipe are cut with an inoculating needle and transferred to a suitable nutritive medium. The mycelium will develop and later, spawn can be obtained. Tissue cultures are much favoured for spawn production because there is no variability and so no sexual spores are involved.

Spawn

The propagating material used by the mushroom growers for planting beds is called spawn. The spawn is equivalent to the vegetative seed of higher plant. Quality of spawn is basic for the successful mushrom cultivation. No matter how carefully all the other requirements for its cultivation are fulfilled.

Mushroom culture preparation basically needs a nutrient medium of malt extract of agar. This can be prepared by taking 1 gm of peptone, 1 gm of malt extract, 0.6 gm of sodium chloride, 1 gm. of agar dissolved in 100 ml water. This makes 100 ml. of malt extract agar. This quantity could be used for preparing 20 tubes of mushrom culture. Take all the materials except agar in 100 ml of water mix them thoroughly in a beaker. Heat the mixture on a low flame constantly stirring the mixture to facilitate dissolving the ingrediants completely. This mixutre is required to be slightly alkaline. For this, adjust the pH in the range of 7.4 to 7.8 by adding a few drops of 0.1 normal sodium hydroxide. Now add 1 gm. of agar to this mixture. Transfer the agar media into 20 clean and dry test tubes at the rate of 5 ml in each and plug with clean sterilised cotton tightly.

Sterilise the tubes for 7 minutes in an autoclave at 15 lbs. pressure per square inch. A pressure cooker can be used in the alternative. Now cool the tubes to room temperature in a slanting position and place the pure culture of mushroom of the desired edible variety into these agar slants duly replacing the sterilised cotton plugs again with fresh sterilized cotton tightly.

Pure culture can also be obtained from the tissue of the fruit body of the mushroom under sterilised conditions by cleaning the surface of the fruit body with 90 per cent alcohol. A small piece of tissue from the sterilized fruit body of mushroom is transferred into the nutrient agar slants (test tubes) under hygienic conditions with the help of a sterilized foreceps.

It is better that these tubes are kept in an incubator for 10 days. In the absence of an incubator, the inoculated tubes may be kept under hygienic conditions at room temperature. The tissue in the test tube grows into mycelium in about a week's time. This is used for inoculation of spawn bottles.

Preparation of spawn

Spawn is a living ramified mycelium multiplied under highly hygienic conditons and used as seed material for cultivation of mushrooms under controlled climatic conditions throughout the year.

Chopped straw or grain of paddy, sorghum, wheat or maize could by used as a base material for multiplying the mycelium in the spawn bottles.

Having secured the above material, adopt the following procedure step by step to prepare spawn.

1. Wash the chopped straw or grain of paddy, sorghum or maize in fresh water.
2. Cook the washed material for thirty minutes, and drain excess water.
3. Mix lime with the grains (@ 20 g per kg of dry wheat in preparing spawns in bottles and 30 g per kg of dry wheat in preparing spawns in polypropylene packets). Add,if there is an apprehension for a bacterial attack, any bacteriostatic

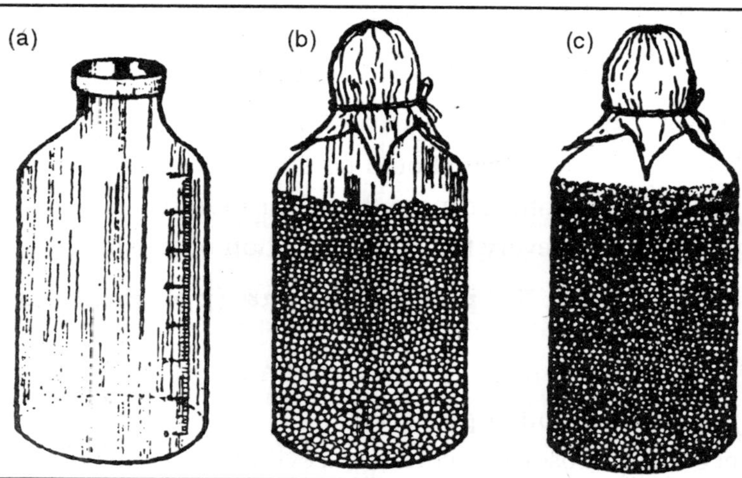

Fig. 6. Preparation of Spawn.
(a) Steriled empty saline bottle, (b) Cooked Sorghum + lime, (c) Autoclaved spawn (sterlized)

compound like chloramphenicol capsule (e.g. enteromycetin capsule 250 mg @ 1 capsule per kg of dry wheat).

4. Fill this mixture upto 3/4ths of the spawn bottle and plug it tightly with sterilized cotton.
5. Cover the mouth of the bottle over the plugged coton with 4" square-clean paper bits and tie it with twine (Fig. 6).
6. Autoclave these bottles for 90 minutes at 15 lbs pressure per square inch and cool them to room temperature.
7. For inoculating the bottles hygienically, a small room may be sterilised by spraying 10 per cent fomaldehyde and closing the room for ten minutes after spraying. The work place should also be sterilised with formal dehyde to avoid any further contamination.
8. Inoculate the sterilised bottles under hygienic conditions in an inoculation chamber under an ultra violet lamp. In the absence of an ultra-violet lamp, the inoculation can be done close to the flame of a spirit lamp.
9. Inoculation is done by placing small bits of inoculum of desired edible variety of mushroom into sterilised spawn bottles duly replacing the sterilized cotton plug again with fresh sterilised cotton plug and covering with pieces of clean paper and tying with twine.
10. Keep the inoculated bottles in cool room with a temperature of 23°±2°C for 10 to 15 days. The room can be kept cool by hanging gunny cloth and keeping it wet.
11. The mycelium will grow in the covered bottles and spawn is ready for cultivating mushrooms on a large scale.

Precautions

1. Inoculum should be free from contamination all through.
2. Persons inoculating should be free from skin diseases and other contagious diseases.
3. In case there is green or black mould develpment in the spawn bottles, the entire contents of the bottles should be burnt.
4. Do not use spawn stored for more than four months.
5. Use only fresh spawn every time for cultivation of mushrooms.

Preparing grain spawn in polypropylene (pp) Packets **(Bhowmik et al., 1992)**

1. Take a pp packet (20 cm by 12 cm) and fill in it 200 g of prepared grains.
2. Put the package in another pp packet.
3. Roll a piece of non-absorbent cotton in a cylindrical form with a length of 5 cm and a circumference of 8 cm.
4. Use the rolled cotton to plug the mouth of the double-packet container.

5. Place the cotton-plug inside at one corner of the mouth and bring the rest with small foldings on the top of the plug.
6. Tie a knot with a cotton thread at 2 cm down the edge of the double packet.

Processing Liquid Medium (Bhowmik et al., 1992)

A. Select some old but solid potatoes of good quality.
B. Take a measure of 750 g of potato and wash them thoroughly in fresh water.
C. Peel off each potato and cut it into 1.5 cm cubes.
D. Wash the cubes lightly in fresh water.
E. Boil 6 litres of water and allow it to cool down.
F. Take out 4 litres of clean water by way of siphoning and leave the rest with sediments at the bottom of the container.
G. Add the potato-cubes in an aluminium container with 2 liters of clean water and boil the contents until the potato-cubes become soft enough but do not split.
H. Take out potato-cubes and allow the liquid to cool down.
I. Add clean water to make up the volume of the liquid to 3 liters and filter the liquid through a thin layer of absorbent cotton.
J. Add 60 g of dextrose to the liquid and stir till dextrose gets thoroughly dissolved.
K. Take in a small container a little amount of the liquid and add to it the contents of 3 capsules of enteromycetin stir the liquid till the contents get thoroughly dissolved. Add the mixture to potato-dextrose medium and shake them to mix up thoroughly.
L. Wash 9 bottles (used saline bottles) and add in each bottle a few broken glass-pieces. Pour 300 ml of the liquid in each of 8 bottles and 600 ml in the 9th bottle.
M. Plug the mouth of each bottle by placing thin layer of non-absorbent cotton between the rubber-cock and the glass-mouth of the bottle. Place a thin layer of cotton-pad on the top of the rubber-cock. Cover the mouth by a piece of thick paper and tie a knot over the paper at the neck of the bottle.

Sterilizing Processed Media

Sterilize the bottles or the packets containing grain medium at a pressure of 15 pounds per square inch (i.e. at a temperature of 121°C) for one hour a day in two consecutive days, or continuously for 2 hours on the same day. Sterilize the bottles containing liquid medium at a pressure of 15 pounds per square inch (i.e., at a temperature of 121°C) once for half an hour.

Ecomonics

In carrying on the production functions as spelt out in the technology schedule, the intending mushrom spawn producer will be required to make an investment of some capital to meet non-recurring as well as recurring expenses. The financial implications will be as follows:

Item of Expenditure	Quantity	Amount (Rs.)
I. *Non-recurring*		
A. Pressure-cooker: 22 litre capacity	1 pc	1600.00
B. Kerosene Stove: 5 litre capcity	1 pc	200.00
C. Aluminium cauldron (dekchi) with lid: 30 litre capacity	1 pc	100.00
D. Aluminum/steel strainer fitted with a handle	1 pc	15.00
E. Sieve with iron wire-net and having mesh smaller than wheat grain	1 pc	25.00
F. Spirit lamp .	1 pc	25.00
G. Syringe (50 ml size) with needle (18 size)	1 set	50.00
H. Bottle fitted with rubber-cock (used saline bottle)	36 pcs	18.00
I. Weighing scale with Weights : 2 kg size	1 set	150.00
J. Small tools and other appliances	LS	227.00
TOTAL (A)		2400.00
II. *Recurring:*		
A. Culture bottle having 300 ml of fresh liquid culture	1 pc	60.00
B. For inoculation of 9 bottles of grain medium		
(*i*) Wheat @Rs. 3.00 per kg	1200 g	3.60
(*ii*) Lime (calcium carbonate) @Rs. 20.00 per kg.	24 g	0.48
(*iii*) Fuel expenses @ Rs. 0.50 per bottle	9 bottles	4.50
(*iv*) Miscellaneous expenses (including costs for disinfectants cotton, etc.,) @ Rs. 0.25 per bottle	9 bottles	2.25
TOTAL (B)		**10.83**
C. For inoculation of 60 packets of grain medium		
(*i*) Wheat @ Rs. 3.00 per kg	8 kg	24.00
(*ii*) Lime (calcium carbonate) @ Rs. 20.00 per kg.	240 kg	4.80
(*iii*) Polypropylene packet (size: 20 cm × 12 cm; and thickness: 80 guage) @ Rs. 0.15 per packet	120 pkts.	18.00
(*iv*) Fuel expenses Rs. 0.20 per packet	60 pkts.	12.00
(*v*) Miscellaneous expenses (including cost for disinfectants, cotton thread, non-absorbent cotton, etc.) @ Rs. 0.10 per packet	60 pkts	6.00
TOTAL (C)		**64.80**

Item of Expenditure	Quantity	Amount (Rs.)
D. For inoculation of 9 bottles of liquid medium and regeneration of 1 bottle of liquid culture consumed		
(*i*) Potato @ Rs. 3.00 per kg	750 g	2.25
(*ii*) Dextrose @ Rs. 35.00 per kg	60 g	2.10
(*iii*) Enteromycetin capsule @ Rs. 0.99 per pc	3 pcs	2.97
(*iv*) Fuel expenses @ Rs. 0.50 per bottle	9 bottles	4.50
(*v*) Miscellaneous expenses (including costs for disinfectants, cotton thread, non-absorbent cotton etc.) @ Rs. 0.25 per bottle	9 bottles	2.25
TOTAL (D)		14.07
TOTAL (III) = A + B + C + D –60.00 + 10.83 + 64.80 + 14.07		149.70

Note:	Expenditure varies from time to time and place to place.

Thus, a person intending to produce mushroom grain spawn is to make a minimum investment of Rs. 2,549.70 with a break-up of Rs. 2400-00 to meet non-recurring expenditure and Rs. 149.70 to incure recurring expenses. The unit with this investment may produce only 9 bottles of grain spawns, 60 packets of grain spawns and 9 bottles of liquid culture as well as regenerate 1 bottle of culture. This may not be sufficient to provide the spawn producer with opportunities for practising the techology learnt or making an optimum use of non-recurring expenditure made. It is, therefore, required to increase the volume of investment by making more provisions for recurring expenses.

It has been assessed that a Small-Size Mushroom Spawn Production Unit following the technology-schedule as given is capable of producing 9 bottles of grain spawns, 690 packets of grain spawns and 9 bottles of liquid culture as well as regenerating 1 bottle of liquid culture by way of utilisation of labour of one person in a month of 25 working days, spending a day for preparation of grain spawns in bottles and one more day for preparation of fresh liquid culture in bottles plus 23 days for preparation of grain spawns in packets. The commercial running of a production unit of this kind requires a working capital to support recurring expenses for at least 3 consecutive months, and hence the investment requirements will be as follows:

Item of Expenditure	Amount (Rs.)
I. *Raw Materials*	
A. Procurement of 1 culture bottle	60.00
B. Preparation of 27 bottles of grain spawns (10.83 × 3)	32.49
C. Preparation of 2070 packets of grain spawns (64.80 60 × 2070)	2235.60

Item of Expenditure	Amount (Rs.)
D. Preparation of 27 bottles of fresh liquid culture and regeneration of 3 bottles of liquid culture consumed (14.07 × 3)	42.21
II. *Labour of 1 person for 75 working days* @ Rs. 30.00 per working day (30.00 × 75)	2250.00
III. *Infrastructural Support* @ Rs. 250.00 per month (250.00 × 3)	750.00
TOTAL (I + II + III)	5370.30

Note:	The expenditure varies from time to time and place to place.

Thus in making this kind of Small-size Mushroom Spawn Production Unit functionally operative and commercially viable, the requirement for capital investment will be Rs. 7770.30, including a capital of Rs. 2400.00 for non-recurring expenditure and Rs. 5370.30 for recurring expenses.

The monthly depreciation on account of non-recurring expenses will be Rs. 40.00, assuming a 5-year life-span for the non-recurring items. Thus, the total monthly expendiure will be Rs. 5370.30/3 + Rs. 40.00 = 1830.10 for producing 699 units of mushroom spawns (27 bottles +2070 packets of grain spawns in 3 months, i.e. 2097 units 3=699 unite per month), each unit having 200 g of grain spawns. This cost per unit of 200 g of grain spawns will be Rs. 1830.10 699 = Rs. 2.62. After taking the management cost, cost for capital investment, contamination loss and storage loss into account, the spawn producer will be able to sellout one unit of grain spawn (200 g) to a mushroom dealer at Rs. 3.50, and the mushroom dealer may sell it to a mushroom grower at Rs. 4.50.

Note:	1. At present the selling rate in the market from Rs. 20 to Rs. 40 per 250 g of spawn. 2. The selling rate depends on the investment.

❑❑❑❑❑

CHAPTER 7

FACILITIES FOR CULTIVATING OYSTER MUSHROOMS

CONTENTS

(*a*) A thatched shed or room preferably with a temperature of 26°± 2°C and a relative humidity of 80 to 95 per cent.

(*b*) Spawn. Generally one spawn bottle containing 300 gms of spawn for two beds of straw.

(*c*) Dried paddy straw (500 gms. per bed).

(*d*) Polythene bags of 70 to 100 gauge. The size of the bags should be 30 cms. wide and 60 cms. long. The bags should have two round holes of 1 cm. diameter in the middle of the bag on each side. 4-5 holes for aeration.

(*e*) Twine enough to tie the mouth of each bag.

(*f*) Racks of the size 6 feet long, 1 foot wide and 5 feet height with 3 shelves. These racks could be either of wood, bamboo and casurina poles for keeping the polythene bags.

(*g*) Thin gunny cloth—Enough to tie as curtains alround the racks to prevent direct light. The gunny cloth will also help keep the room temperature cool and the desired humidity when wetted enough.

(*h*) Iron rod of 1 foot length for removing the spawn.

(*i*) Plastic trays of size 20" long 15" wide and 2" height for mixing of the spawn thoroughly before filling the polythene bags.

(*j*) Dettol antiseptic lotion—Mix 2 Teasopoon full dettol with 1 litre clean water and use this to clean hands before touching any material in the thatched shed or room growing mushrooms.

(*k*) A foot bath with dettol anticeptic lotion at the entrance of each shed. Dip feet in the foot bath before entering the thatched shed room.

(*1*) BHC 10 per cent dust for sprinkling alround in and outside the shed and along the walls to prevent ants.

Procedure

(*i*) Chop the paddy straw into small bits of 3 to 5 cms and soak them in clean drinking water for 12 to 16 hours in a suitable container.

(*ii*) Drain of the excess water after 12 to 16 hours.

(*iii*) Immerse these small straw bits into boiling water for 15 to 30 minutes for sterilization.

(*iv*) Take out the paddy straw bits and dry in shade for about 30 minutes.

(*v*) Wash the iron rod, tray and hands with clean water and rinse with dettol solution.

(*vi*) Clean spawn bottle from outside with dettol solution.

(*vii*) Remove spawn from the bottle into the tray and mix well with hands.

(*viii*) Divide the well mixed spawn in the tray into two portions.

(*ix*) Take polythene bags and tie them at the bottom with twine.

(*x*) Fill each polythene bag to about 5 cms. of the paddy straw bits to form the first layer.

(*xi*) Take handful of spawn and sprinkle over the entire straw layer and cover again with straw bits to another 5 cms.

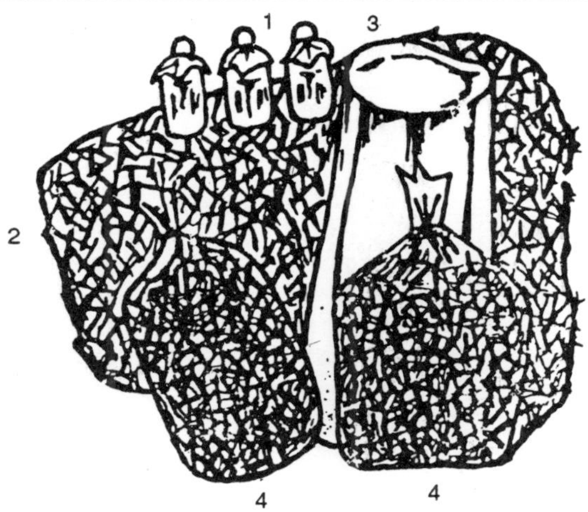

Fig. 7. Preparation of bed.
(1) Spawn Bottles, (2) Sterlized Paddy Straw, (3) Polythene bag (4) Bed

(*xii*) Repeat alternatively filling the spawn and straw bits three times and tie the mouth of the polythene bag with twine (Fig. 7), known as bed.

(*xiii*) Clean the racks with clean water and place the filled polythene bags vertically on the shelves of the rack one beside the other leaving a space of about six inches between the bags (Fig. 8).

Fig. 8. Beds are arranged on racks.

(*xiv*) Leave the racks with the polythene bags for about 15 to 18 days.

(*xv*) Check humidity and temperature and sprinkle water to maintain the required temperature of 26 ± 2°C and humidity of 80-95%.

(*xvi*) Keep the room preferably clean, cool and dark. Spray water on the floor and gunny if needed.

(*xvii*) Check the bags periodically say once in 3 days and remove the contaminated bags *i.e.* those with green or black mould growth. Take care to maintain clean and aseptic conditions each time while entering the thatched shed or room.

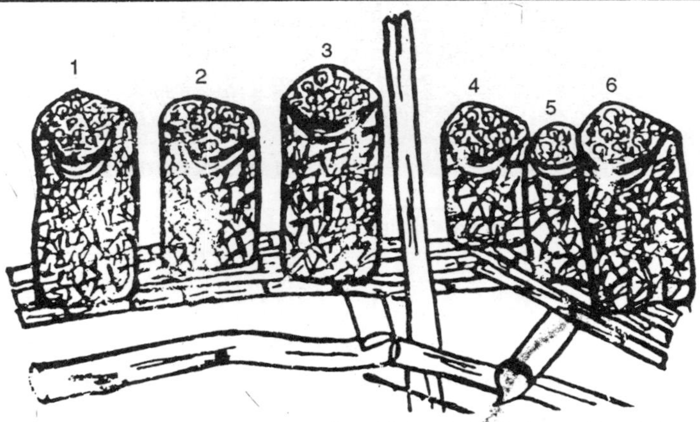

Fig. 9. After removal of polythene cover.

(*xviii*) After about 12 to 15 days, remove the polythene covers and keep the growing mushroom lenght wise on the rack with least disturbance to the growing mushroom (Fig. 9).

(*xix*) Avoid watering beds for first two days.

(*xx*) After two days, sprinkle water on beds twice or thrice daily.

(*xxi*) Allow aeration and light into the shed for about 30 minutes every morning and evening by opening windows and taking gunny cloth up.

(*xxii*) Harvest first flush of mushrooms after four to six days by cutting fully grown mushroms at the base of the stem of the mushroom. Take care to see that aseptic conditions are maintained.

(*xxiii*) Two to three harvests can be had at an interval of 7 to 10 days between each harvest [Fig. 9*a*].

(*xxiv*) After mushrooms are harvested, the paddy straw beds could be used as manure for other crops. Remove them to a compost pit far away from the thatched shed.

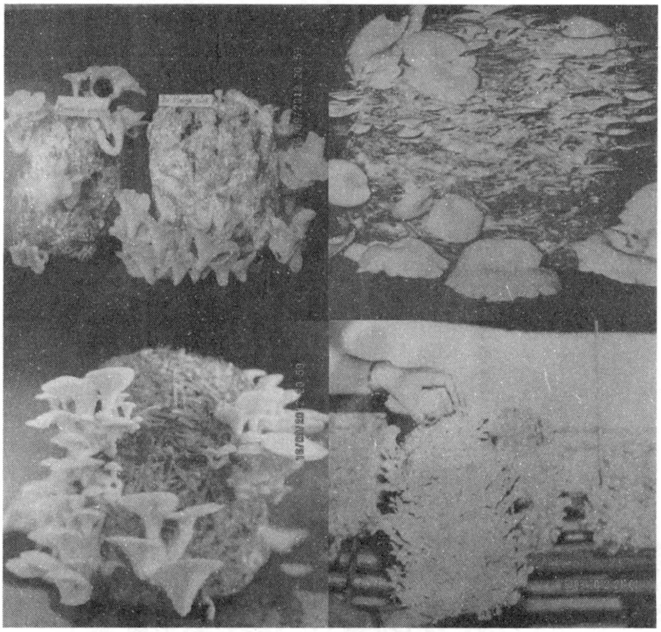

Fig. 9a. Oyster Mushrooms

Yield

Each polythene bag with spawned paddy straw on an average gives about 1 to 2 kg. of fresh mushrooms. This means that from every two bags of material in the polythene bags it yields 2 to 4 kg. of mushrooms. (Vijayakhader, 1988)

Chemical sterilization

In India oyster mushroom is generally cultivated on the steam pasteurized wheat or paddy straw (Jandaik, 1976). This method gives satisfactory performace as long as fungal contaminants viz. *Trichoderma viride, Cladosporium, cladosporioides, Doratomyces stetnonitis, Mucor spp. and Rhizopus spp.* do not invade the substrate (Vijay & Sohi, 1987; Vijay et al., 1985; Vijay & Sohi, 1989). Further, the method is costly and cumbersome besides being a limiting factor in the cultivation of this mushroom on a large scale. Vijay & Sohi (1987), in their pursuit to recycle agriculture wastes for the cultivaton of this delicious mushroom, developed a simple chemical sterilization technique (CST) of the substrate. The method involves the steeping of straw in Formalin (40% formaldehyde) and Bavistin (50% Carbendazim) solution (500 + 75 ppm) for 18 h followed by spawning. This method has since been giving satisfactory performance and is being successfully adopted by the growers.

Higher concentrations of Formalin may be retarding the mycelial growth as observed by Rajarathnam et. al. (1983). With urea, formalin is reported to form a sugar(Rajarathnametal., 1983) and thereby neutralising its direct effects.

In some bags where treated straw was washed with water before spawning, incidence *of Mucor sp.* was noticed. Formalin present in the straw at spawning is not at all harmful and is rather beneficial to the crop and provides protection against Mucor infection. Further it was found that same chemical solufions can be used for three dippings.

Effect of he following spawn bases viz., (i) saw dust + rice bran (8 : 1), (ii) saw dust 20% + cotton waste 20% + rice bran 5% + calcium carbonate 5%; (iii) cotton waste 90% + maize meal 15% + calcium carbonate 5% + sugar 5%; (iv) saw dust 80% + rice bran 20% + calcium carbonate 1 % + maize meal 2%; (v) sorghum grain + peptone 3% + sugar 5% + calcium carbonate 2%; (vi) sorghum grain + sugar 5% + calcium carbonate 2%; (vii) rice straw spawn with 15 g horsegram and (viii) sorghum grain with 2% calcium carbonate on the sporophore production *of P. sajor-caju* was assessed and found good.

Economics of a medium size oyster mushroom farm taking 20 crops per annum. (Prakash & Tejaswini 1991)

A. Non-recurring Expenditure	Rs.
1. Drums with Heating coil (2 Nos.)	500
2. Chaff cutter	1,000
3. Wooden table	200
4. Wire mesh frame	100
5. Bamboo racks	1,000
6. Sprayer & Buckets	800
7. Drums for soaking straw (5 Nos.)	1,000
	4,600
Interest @ 15% on Rs. 4,600	690
Depreciation @ 20% on item 1-7 :	920
Total	1,610

B. Recurring Expenditure	
1. Paddy straw, 16 tonnes	11,200
2. Fungicides and insecticides and other chemicals	500
3. Spawn bottles (excluding the cost of bottles)	8,000
4. Polythene/Polypropylene bags	6,500
5. Labour permanent-1	6,000
6. Casual Labourer-2	7,000
7. Electricity and Water etc.	3,500
8. Miscellaneous	2,000
9. Rent for building	4,800
Total	49,500 or 50,000
Anticipated yield	4,800 kg
(Sale @ Rs. 15 per kg.)	Rs. 72,000
(Sale @ Rs. 30/- per kg.	Rs. 1,44,000
Total Expenditure Rs. 50,000+1,610	= Rs. 51,610
Net income Rs. 72,000—51,610	= Rs. 20,390

If the sale price is Rs. 30 per kg net income is = 1,44,000 - 51,610 = 92,390. At present sale price ranges from Rs. 25 to Rs. 35 per kg.

Note:	Expenditure as well as sale price veries from time to time, place to place.

❑❑❑❑❑

CHAPTER 8

FACILITIES FOR CULTIVATING WHITE BUTTON MUSHROOMS

CONTENTS

The white button mushroom (*Agaricus bisporus*) is very popular throughout the world and is the most important mushroom of commerce. It can be successfully cultivated in places where the environmental conditions are favourable. The optimum temperature for mycelial growth is 22-25°C and that for fruit body formation, 14-18°C. Due to its low temperature requirement, its cultivation is restricted to the cool climatic areas and to the winter in the plains of Northern India. This mushroom also needs a high percentage of relative humidity. The substrate for cultivation is a specially prepared compost. The mushroom house should have facilities for temperature control and pasteurisation. Inside the house, shelf system is usually adopted for cultivation. Buildings are constructed of wood or hollow cement bricks or double walls, in Taiwan special houses of 20 m long and upto 6 to 7 m wide and 4 to 5 m height are considered to be ideal. The shed is portioned into small compartments and provided with trays. Environmetal conditions like temperature, relative humidity and ventilation are controlled inside the shed by installing suitable equipment.

Spawn

The propagating material used by the mushroom growers for planting beds is called spawn. The spawn is equivalent to the vegetative seed of higher plant. Quality of the spawn is basic for the successful mushroom cultivation.

In the early days, the soil was dug up from the places where the mushrooms were known to grow and used for planting new beds. After the beds were fully covered with spwan, portions were used dried or fresh, to inoculated other beds. The spawn was called 'flake spawn'. Later 'brick spawn' came into use. It was made by mixing composted horse-manure with loam soil and made into bricks or cakes 5 cm thick. A hole was made in one edge and inoculated with manure containing the mushroom mycelium. The bricks were then left to dry. Since pure culture was not used for making flake and brick spawns, molds and pests appear in the newly planted beds.

The pure culture spawn was first introduced in 1905 in the U.S.A. It entirely replaced the old spawn types then in use. At present, the pure culture spawn has been the basis of modern spawn production units all over the world. The manufacure of the pure culture spawn is done under scientifically controlled conditions which demand a standard of hygiene as in a hospital operation theatre. Equipment and substrate used for spawn are autoclaved and filtered air is passed during the inoculation esures complete freedom from contamination.

Previously, composted horse-dung or tobacco stems were used as substrate. Dung-spawn was in the form of cake which was broken up by the grower and distributed in the compost. In 1931, Sinden introduced grain as a substrate. The mycelium grows around individual grains thus allowing a better distribution of inoculum in the compost. Manure spawn is, howeve, preferred for poorly prepared compost. However, with the grain-spawn, mycelium grows mainly in the surface of the grains and may be killed by even small amounts of ammonia in the compost. The mycelium grows in the interior of the lumps of manure spawn and it may survive to grow out at a later stage when residual ammonia has evaporated.

(*a*) ***Manure Spawn.*** Both composted horse-dung or synthetic compost may be used. The composted manure is thoroughly washed to remove such substances in compost which retard growth. The excess water is squeezed out the moisture content adjusted to 60%. The manure is packed in half-litre milk bottles or heat-resistant polypropylene bags of suitable size. The bottles or bags are plugged with non-absorbant cotton-wool and sterilized in an autoclave at 121°C for 2 hour on 2 consecutive days for an hour each. They are then inoculated with a large bit of agar-containing mycelium and incubated at 22°-24°C in a dark place. The spawn will be ready in about 2 weeks. The prepared spawn can be sued to inoculate fresh bottles or bags to obtain the second generation spawn.

(*b*) ***Grain Spawn.*** Ten kilograms of wheat grains are boiled for 15 min. in 15 litres of water and then allowed to soak for another 15 min. without heating. The excess water is drained off and the grains are cooled in sieves. Turn the grains several times with a spoon for quicker cooling. The cooled grains, are mixed with 120 g of gypsum

($CaSO_4.2H_2O$) and 30 g of calcium carobonate. The gypsum prevents the grains from sticking together and calcium carbonate is necessary to correct the pH. The prepared grains are filled into half-litre milk bottles or polypropylene bags (at the rate of 150-200 g per bottle or bag) and autocalved for 2 hour at 121°C. After sterilization, the material should have a pH value of 6.5 to 6.7. The bottles are inoculated with grain spawn or with bits of agar medium colonized with mycelium and incubated at 22°-24°C in a dark place. The mycelium completely permeates the grains in about 2 weeks. Other grains like sorghum and pearlmillet can also be used for spwan making (Fig. 6).

Compost. The white-button mushroom is grown on select substrate which provides adequate levels of nutrients to support the crop so that it can successfully compete with other micro-organisms. Traditionally, partially decomposed horse-manure has been the principal medium for providing the required nutrients in artificial cultivation of the mushroom and it is only in recent times that other materials have also been used successfully.

(a) Materials and their functions

(*i*) ***Base materials.*** These include wheat straw, maize cobs and other similar cellulosic plant wastes with or without horse-manure. Conventionally wheat straw either alone or mixed with horse-manure is the most widely used base material. When wheat straw is not available, straws of other cereals, like rice or barley may be used. The chief function is to provide cellulose, hemicellulose and lignin in bulk. These materials also provide proper physical structure to the mixture to ensure the necessary aeration for the build up of microbial population and the subsequent spawn growth in the compost. Rice and barley straws are quite soft and decompose quickly, leaving only a little fibre for imparting a proper physical structure to the compost. Therefore the type and quantity of supplement should be discretely utilized at the proper time.

(*ii*) **Supplements.** These are for activiating fermentation and can be categorised as:

Animal dungs. These include horse- and chicken-manure, the extremely variable manures in composition. Nitrogen content may vary from 1 to almost 5%. In addition to nutrients, they contribute greatly to the final bulk density of the compost. Cow manure is not considered suitable.

Carbohydrate nutrients. From molasses, wet brewers' grain and malt sprouts, carbohydrates are readily available.

Concentrate meals. These materials are usually used for animal feeds and include wheat or rice bran, dried brewer's grain, the seed meals of cotton, soya, castor and linseed. In these, both nitrogen and carbohydrate are available rather slowly. Nitrogen content may vary from 3-12%. The oil and mineral content of some of these may be of significance in mushroom nutrition.

Nitrogen fertilizers. Nitrogen in chemical fertilizers (ammonium sulphate, calcium ammonium nitrate and urea) is rapidly released for the quick growth of microbia population.

Materials to correct mineral deficiencies. These are muriate of potash and calcium super phosphate.

Materials to correct greasiness. Gypsum and calcium carbonate serve to precipitate suspended colloidal materials and neutralize greasiness.

The choice of materials within each category is largely determined by cost-factors and their availability locally. Composts prepared from horse-dung mixed with straw are termed as 'natural', whereas they are called synthetic if the base material used is mainly straw without bulk animal manure.

(b) Wheat Straw

Wheat straw was first advocated by the Polish scientist Treschow (1944), who discovered that it was possible to obtain heavy growth of spawn threads on xylose, a form of carbohydrate present as xylan in wheat straw. Straw protected from rain is preferred. One-year-old straw which is no longer bright yellow and shiny, can be used only if it is tough. Full-length straw must be chopped to smaller size, about 8-10 cm length, or else the heap would be less compact. Such a heap would not be able to retain moisture and fermentation would be slower. The reverse, if the straw is too short, the heap would be compact and with very little air space inside allowing anaerobic fermentation. Straw, as is sold in the market for cattle feed is quite suitable. Composting is a microbial process requiring biological changes in all parts of the straw.

(c) Horse-manure

Stable manure with wheat, barley and hay-bedding must be collected regularly from the stables at intervals not more than a fortnight and fermet

(d) Composting Theory

Composting for mushrom cultivation has three basic purposes: (i) To convert more suitable substrate, (ii) To create a favourable medium (iii) its fermenting temperature is high enough to elminate most harmful pests and diseases.

Compost and Compost Preparation

The quality of compost influences the yield of mushrooms to a great extent. This in turn depends on the raw materials for composting and also on the process of composting followed. The compost is obtained by microbial degradation of organic wastes. These wastes include vegetable and animal matter, forest litter, straw, animal manure etc.

Compost prepared out of horse manure and wheat straw is the ideal one. Since these materials are not easily available everywhere, many substitutes are suggested and are in use. In recent times mushroom cultivation has become very popular in Taiwan, where techniques are perfected to prepare the compost using paddy straw.

There are two methods of composting, long method and short method. This distinction is based on the time taken for composting and the long method needs three to four weeks, while the short method requires only 12 to 15 days, since composting process is hastened by pasteurisation. In the long method pasteurisation is avoided, which will make the compost poor in quality and often gives variable yields. Because of all these reasons, the long method of compost making is considered to be a less efficient system.

Long Method

The composting is done on a cement floor. It can be done in the open or under a roof, but the sides are to be kept open.

Natural Compost

In many parts of Europe, the compost is traditionally prepared by using the horse-manure and the bran waste consisting mainly of straw bedding of wheat or barley. These-are taken in the proportion 4:3 and to this mixture, generally 100 kg substrate mix. The manure in kept as a heap of about one metre height. The heap is regularly examined and when it emits a strong smell of ammonia it is turned down repeatedly. This is to be done in every three or four days. Finally to every tonne, 25 kg. of gypsum is also added.

Synthetic Compost

Many formulae are available and the ingredients to be used varies according to the locality and availability of materials. A widely followed formula in India, is the one recommended by Kapoor and Nita Bahl (1982) from Indian Agricultural Research Institute, New Delhi.

1.	Wheat straw (chopped to 8-20 cm long)	250 kg
2.	Wheat bran	25 kg
3.	Ammonium sulphate or Calcium ammonium nitrate	4 kg
4.	Urea	3 kg
5.	Gypsum	20 kg

Chopped wheat straw is spread over the floor and water sprinkled over the same so as to wet it thoroughly. The other ingredients except gypsum are mixed thoroughly with the wet straw. Finally the mixture is stacked well to a height of one metre and compacted. This mixing and heaping can be easily done with the help of wooden boards. The board prepared properly are placed in position and the mixed substrate put inside and slightly compressed. This mixture is turned periodically in accordance with the schedule on every 5th, 10th, 14th, 18th, 22nd and 26th day. The gypsum is added in two equal splits on the 14th and on the 18th day. It is also advisable to add a nematicide like nomagon on the 22nd day (40-50 ml per tonne of substrate). In Mushroom houses, where insect pests are also a problem, ten to fifteen ml of malathion is also added during the final turning and mixed with water. Sufficient quantity of water is also sprinkled on the heap. When the compost is ready it will be a dark brown mixture without the smell of ammonia.

Some of the alternative formulae available are given below: Indian Institute of Horticultural Research, Bangalore. Long method—Schedule of turning—0, 6, 10, 13, 16, 19, 22, 25 and 26th days (filling)

1.	Wheat straw (cut into pieces)	250 kg
	or Paddy straw	400 kg
	Ammonium sulphate or Calcium ammonium nitrate	9 kg
	Super phosphate	9 kg
	Urea	4 kg
	Wheat bran	30 kg
	Gypsum	12 kg
	Calcium carbonate	10 kg
Note.	It is advisable to add 6 kg of cotton seed meal also when paddy straw is used.	
2.	Paddy straw	150 kg
	Maize Stalk	150 kg
	Ammonium sulphate	9 kg
	Super phosphate	9 kg
	Urea	4 Kg
	Rice bran	50 kg
	Gypsum	12 kg
	Calcium carbonate	10 kg
	Cotton seed meal	5 kg
3.	Paddy straw	3 ton
	Chicken manure	1.5 ton
	Wheat bran	125 kg
	Gypsum	90 kg
	Formulae sugested by Seth (1976)	
1.	Wheat straw	300 kg
	Calcium ammonium nitrate	9 kg
	Urea	3 kg
	Super phosphate	3 kg
	Muriate of potash	3 kg
	Wheat bran	15 kg
	Molasses	5 kg
	Gypsum	30 kg
	B.H.C. 5%	250 kg
	Nemagon	30 ml
2.	Wheat straw	300 kg
	Poultry manure	60 kg
	Calcium ammonium nitrate	6 kg
	Urea	2 kg

Super phosphate	3 kg
Muriate of potash	3 kg
Wheat bran	15 kg
Gypsum	30 kg
B.H.C. (50%)	250kg
Nemagon	30 ml
In place of wheat straw, paddy straw (400 kg) can also be used.	
Short Method Formula	
Chopped wheat straw	1000 kg
Chicken manure	400 kg
Barley	72 kg
Urea	14.5 kg
Gypsum	30 kg

Phase I. Outdoor composting

Barley and chicken manure are added to wheat straw and stacked after adding sufficient moisture to completly wet the same. The stack can be 3.3 x 2.5 x 1 m. Turnings are given on the 2, 4, 6 and 8th days. The pH to be adjusted to 8.5 and the compost filled into trays for pasteurisation.

Phase-II

In phase II, pasteurisation is done by dry heat or by steam or by fumigation with methyl bromide.

Steam pasteurisation

Steam or dry heat is introduced to establish an aerobic fermentation and the temperature maintained between 52 to 60°C inside the compost. Usually it is done in a room well insulated where the trays are properly kept and after this all ventilators are closed and steam introduced to raise the temperature to 52 to 54°C, once this temperature is attained it is maintained for 2 or 4 days. After that the temperature of compost is further raised to 58 to 62°C for four hours and afterwards fresh air is introduced and steam supply cut off.

There are also mechanical composting processes utilising special types of barrels.

Spawning

Generally grain spawn is used for cultivation. The compost when ready, is taken in trays leaving about 5 cm from the top. The spawn is then spread evenly and lightly pressed and afterwards a layer of compost also is spread over it. One bottle of spawn is sufficient for three trays. Instead of spread spawn with the compost can also be carried out. After spawn is spread, cover with sterilised news papers. Every day the trays are checked for sufficient moisture and the necessary wetness is maintained by periodical sprinkling of water. The relative humidity inside the mushroom house during the growing season is to be maintained at 90 to 56%. Spawn run will be optimum at a temperature of 25 ± 2°C. If conditions are congenial, mycelium starts spreading and permeates through the compost. This stage is called the 'spawn run stage' and will be complete in 15 to 20 days. Once the spawn run is complete the news Papers can be removed.

Characters of a Good Spawn

(*i*) Spawn prepared on Jowar (sorghum tricolor) or wheat grain gives the maximum yield over the spawn prepared on bajara, barley or kodo grains.

(*ii*) There should be a proper coating of the mycelium around every grain used as a substrate for spawn prodction. No loose grains should be seen in the bottles when these are bound together with the mycelium. The grains left over without mycelial coating will invite contamination in the compost during spawn running period.

(*iii*) The growth of the mycelium in the spawn bottles should be of silky/stand type. It should be cottony type growth because there is likelihood of stroma formation on the casing layer, which interferes with air exchange and absorption of water in the casing material resulting in low yields.

(*iv*) The growth of fresh spawn is more or less white. Brown colouration develops as spawn grows. Fresh spawn gives higher yield than the old ones. Spawn should not be more than one month old in any case.

(*v*) There should not be any slimy growth in spawn bottles which is an indication of bacterial contamination,

(*vi*) There should not be any greenish or blackish spot in the spawn bottles. Such type of spots indicate that the spawn is contaminated.

Casing

Once spawn run in the compost in the tray is completed, the trays are to be covered with a thin layer (2-5 cm) of soil, after removing the news papers. This is known as casing. It helps to retain moisture and help in gaseous exchange. It also induces fruiting as the vegetative mycelium enter into the mdium which is deficient in nutrition. After casing the temperature is to be maintained at 23-24°C for 8 to 10 days and relative humidity 85 to 90 per cent. Fresh air has to be introduced and in another 12 to 16 days pin heads will start to appear.

Many materials are in use for casing, but the commonly used materals are:

1. Spent compost, sand and lime (4:1:1)
2. Fam Yard Manure (F.Y.M) and Loam (1:1)
3. Soil and sand (1:1)

In the casing medium chalk or calcium carbonate is added to maintain the pH aroud 7.5. All casing materials used are to be pasteurised before use. The material with 20 per cent moisture are taken inside small barrels which are provided with pipes or passing steam. Pasteurisaton can also be done with chemicals like 5% formalin.

Casing

The application of casing layer to spawn run compost induces the vegetative phase (mycelial) to reproduction (fruiting phase.). Casing is a top-dressing applied to the spawn-run compost on which the mushrooms eventually form. F.Y.M. + Loam soil or F.Y.M. + spent compost can be used as casing. Casing does not need nutrients since casing acts as a water reservoir and a place where rhizomorphous form. Rhizomorphous look like thick strings and form when very fine mycelium fuses together. Mushroom initials, primordia or pins form on the rhizomorphous, so without rhizomorphous there will be no mushrooms. Casing should be sterilized to eliminate any insects and pethogens it may be carring. Also it is important that the casing be evenly distributed so the depth is uniform over the surface of the compost. Such uniformity allows the spawn to move into and through the casing at the same race, and ultimately for mushrooms to develop at the same time. Casing should be able to hold moisture since moisture is essential for the development of a firm mushroom.

Managing the crop's casing requires that the compost temperature be kept at around 24°C for up to 5 days after casing, and the relative humidity should be high. There after, the compost temperature should be lowered until small mushroom initials (pins) have formed. Throughout the period following casing, water must be applied intermittently to raise the moisture level to field capacity before the mushroom pins form. Knowing when, how, and how much water to apply to casing is an "art form" which readily separates experienced growers from beginners.

Why casing is necessary

Casing of the mushroom beds or compost is necessary because:

1. Casing soil is a nutrient deficient medium, which helps in converting the vegetative phase into fruiting.
2. Fruiting bodies are formed in abundance and thus production is economical.
3. It helps in the conservation of the environment in mushroom beds.

Characteristics of a good casing

A good material should possess the following characteristics.

(*i*) Good water holding capacity and more pore space percentage.

(*ii*) Capable to release harmful gases during cropping.

(*iii*) Free from harmful microorganisms.

(*iv*) pH should be slightly alkaline.

(*v*) Should be properly decomposed.

Cropping

Pinheads start appearing in the trays in 12 to 15 days after casing and the cropping continues for another two or three months. Regular periodical watering is the most important item of work during this period and watering is to be done carefully by gentle sprayings. There must be provision for free air movement also inside the mushroom house. Usually the buttons are collected every alternate days, when they are about 2 to 5 cm diameter. They are harvested by gently twisting with least disturbance of casing soil. Harvested mushrooms are gently cleaned and packed in thin polythene bags for marketing. In recent days instead of trays, polythene bags are also in use for cultivation.

Note:	For the construction of a mushroom house, with air conditioner, steam pasteurisation system etc., an amount of Rs. 1 lakh is needed for the non-recurring items. The cost of production per kg may vary from 15 to 30 rupees according to locality.

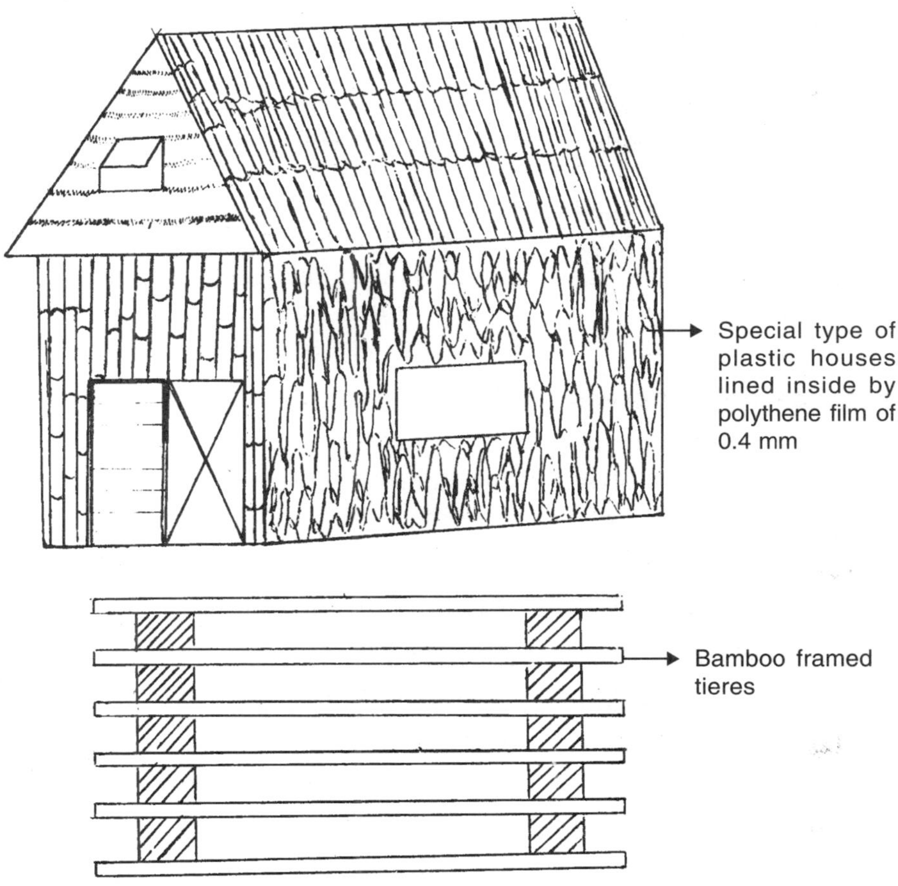

Fig. 10. Cropping House, with bamboo framed tiers

In Asia, in the recent past, Taiwan has made much progress in the production and export of mushrooms, including *Agaricus bisporus.* The cultivation of the same is mainly carried out in special type of plastic houses lined inside by polythene film of 0.4 mm. Inside of the hut is provide with bamboo framed tiers, and the shelves run the entire length (Fig. 10). Typical sheds described by Ho et al. (1968) have given shelves of about 15 m of bed space and 11.38 × 5.18 × 3.66 m of air space. Between bed layers there is a space of approximately 0.55 m and a space of 12.7 cm left below the buttom bed. A foot path of 0.6 m width is left between the two tiers and between side walls and edge of beds for movement. The bamboo used is often pre-treated with preservatives.

The synthetic compost commonly used is prepared as follows:

	(Composition kg/m^2)
Paddy straw	30.3
Ammonium sulphate	0.6
Superphosphate	0.6
Urea	0.15
Calcium	0.9

The straw has to be cut into bits of 15-20 cm, soaked or sprayed with a solution of 2 per cent calcium carbonate. The urea and water are added and the material properly stored. They are turned periodically and at least three turnings are to be given in 3 or 4 days intervals. The ammonium sulphate and superphosphate are added then and turned. After this the compost is pastcrurised by passing steam raising the temperature to 60°C. Spawnsing is done at the rate of 2-3 bottles/3.3 m, and casing is done by using clay loam and calcium carbonate mixture, temperature is to be lowered to 25°C. The requirement for fresh air for *A. bitorquis* is less than that for *A. bisporus* and the quantity of air let in is to be adjusted according to the size of the flush. Water is to be periodically and regularly sprayed.

As the time needed for the button to open is very short in this fungus, the mushroom is to be picked up at the proper time.

Dos and donots of button mushroom growing

(1) Take full precautions while making the compost. There should not be any ammoina smell when the compost is ready.

(2) The compost should not be too wet or too dry at the time of filling. If it is dry then sprinkle water over the compost before filling the boxes.

(3) The success of mushroom growing depends on the compost.

(4) Boxes should be filled up to the brim with the compost.

(5) Spawn should be from a reliable source.

(6) Do not use contaminated spawn.

(7) After spawning cover the boxes with newspaper. Watering should be avoided during the spawn run. If watering is essential, spray water over the newspaper with a fine jet.

(8) The casing soil should be properly sterilised Do not sterilise the casing soil long before the use and store it for a longer period.

(9) Casing soil should not be very fine otherwise there will not be a good gaseous exchange.

(10) Before putting the casing over the spawn run compost, moisten it with water.

(11) Cow dung used for the casing should be rotten and should be one year old.

(12) Maintain the temperature of the room for 2 or 3 days at 22 to 24°C after casing and then lower the temperature to 14 to 18°C.

(13) To keep the room humid and the temperature low, spray the walls and floor of the mushroom house with water.

(14) If the weather is dry hang moist gunny bags on the side of the trays and keep it moist or fill the gunny bags with saw dust and spray water over it to create humidity in the room.

(15) If there is pinning of the casing soil, rake the csing soil gently with a fork or with a pointed knife to provide the gas exchange.

(16) If after raking there is no pinning it is better to remove the whole casing soil and replace the bed.

(17) If there is an appearance of disease in the box, treat the box immediately. If it does not respond to the treatment remove the diseased tray immediately from the room.

(18) Spray water gently, do not water the tray before picking.

(19) After picking fill up the holes with sterilised casing soil and spray water.

(20) For picking, if the mushroom is cut from the base then remove the remaining portion with the pointed knife. Do not leave it as such, otherwise other saprophytes and bacteria will attack it and will be the source of disease.

(21) Keep the butts in one container and burn it or throw it in the pit far away from the mushroom house.

(22) Never broom the mushroom house, dust and spores and of micro organisms will spread and will spoil the mushrom. Remove the dust by sweeping it gently with a duster.

(23) Before entering the mushroom house remove your shoes and put on the slippers kept separately in the mushroom house.

CHAPTER 9

FACILITIES FOR CULTIVATING PADDY STRAW MUSHROOMS

CONTENTS

Facilities Needed for Cultivating Paddy Straw Mushrooms

Paddy Straw mushrooms grow well in temperatures ranging between 20° to 45° C. However, good production can be achieved in an environment of 30°C.

While most of the facilities like a thatched shed or room, dried paddy straw, twine, racks, gunny cloth, dettol, footbath, BHC 10% dust and the like are common, the difference is that the thatched shed or room should be better ventilated and the paddy straw should not be crumpled and the leafy material should be removed before use and the straw be clean. Also needed are large polythene sheets to cover the straw beds, 45 to 50 kg fresh, clean paddy straw are needed to make a bed.

Procedure

(*i*) Tie paddy straw into small bundles of about 10" diameter and 2 feet long.

(*ii*) Take care to see that each bundle weights about 1 to 1½ kg. and is tied well with twine at both ends.

(*iii*) Cut the uneven ends with the help of a big scissors or a small secature.

(*iv*) Soak paddy straw bundles in clean water for about 12 to 16 hours.

(*v*) When the bundles are fully soaked in water, take them out and remove excess water by holding the bundles vertically for about 15 to 20 minutes.

(*vi*) Clean each rack with dettol solution.

(*vii*) Place eight bundles of straw side by side lengthwise on the rack shelf made of bamboo or casurinapoles (Fig. 11).

(*viii*) Take care to see that the bases of the bundles are alternately placed from the edge of the shelves on each rack.

(*ix*) Paddy straw beds thus formed are ready to receive the spawn.

(*x*) Place small portions of spawn 4 to 5 inches inside the edge of the paddy straw bed.

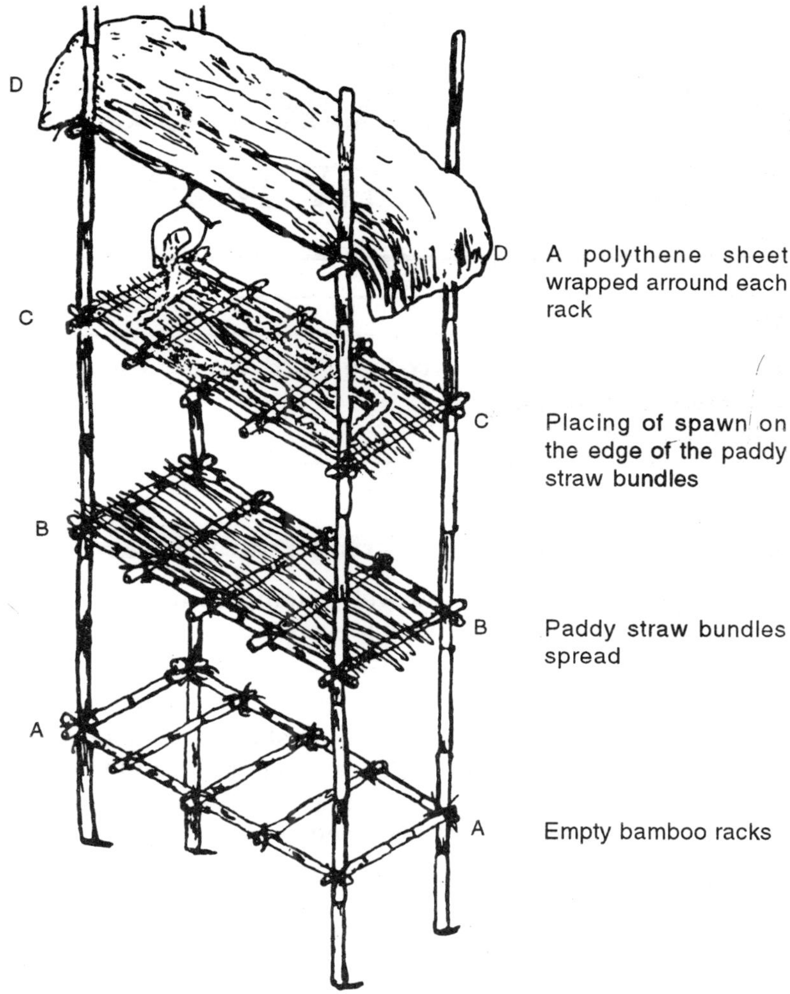

Fig. 11. Paddy straw arranged on bamboo racks.

(*xi*) Sprinkle 2 to 4 gms of finely grounded greengram flour over the small Portion of the spawn placed in the straw bundle to enhance the yield.

(*xii*) Place the second layer of the paddy straw bundles with the spawn and the greengram flour longitudinally to the first row of paddy straw bundles and continue to place one layer over other.

(*xiii*) Firmly press down the layers of paddy straw bundles thus placed over each shelf of the rack and sprinkle water on the bundles in the racks once a day just to keep them moist always.

(*xiv*) A polythene sheet may be wrapped around each rack to reduce evaporation.

Mushrooms appear between 13 to 15 days of spawning and continue to grow for 4 to 5 days. Harvest them carefully. A second crop may appear after one week. Harvest this crop and remove the paddy straw beds to a compost pit. Care must be taken that the most hygienic conditions are scrupulously followed throughout.

Precautions

Follow the precautions as advised earlier for the Oyster mushrooms. Do not enter the thatched shed or room without dipping feet in the foot bath.

Do not forget the steps for maintaining the hygienic conditions.

Use dettol antiseptic lotion to clean hands always before proceeding to do any thing in the shed or room.

Sprinkle profusely BHC dust all along the walls and on the floor to prevent ants.

Harvesting Mushrooms

Small buttons of mushrooms are picked twice a day. Pick only those heads which have grown fully and before they open up and turn into loose umbrella shapes (as in Fig. 12). Do not pick up too small ones as (as in Fig. 12) they are not ready for consumption.

Fig. 12. Different stages of growing paddy straw mushroom

Dos and Donts

1. Do know the characteristics of the species of mushrooms you are growing.
2. Do not consume any other fruiting body which differs from the species being grown.
3. Do not consume decayed mushrooms.
4. Do not use those species whose flesh changes colour when bruised or cut.
5. Do not consume mushroms which exudate milky juice when cut.
6. Do not pick immature fruiting bodies because at this young stage they do not clearly indicate all the characteristics of the species to which they belong.

Economics of paddy straw mushroom on a small seasonal farm taking 3 crops per annum

(Prakash and Tejaswini, 1991)

1.	Paddy Straw 3 tons and supplement	Rs. 2,200.00
2.	Spawn bottles 180 (including the cost of the bottles) @ Rs. 2 each	Rs. 360.00
3.	Fungicide and insecticides	Rs. 50.00
4.	Labour charges	Rs. 500.00
5.	Water, Electricity, Firewood charges	Rs. 300.00
6.	Miscellaneous charges	Rs. 200.00
7.	Rent for Building	Rs. 400.00
	Total	Rs 4, 010.00
	Yield anticipated @ 12% on dry weight	360 Kgs.
	Income @ Rs. 15 per kg.	Rs. 5400.00
	Net Income Rs. 5,400.00 - Rs. 4,010.00	Rs. 1,390.00

Note.	
	1. Spawn bottle rate ranges between Rs. 8-12. Sale price of fresh mushrooms ranges between Rs. 25-40.
	2. Expenditure and sale price varies time to time and place to place.

□□□□□

CHAPTER **10**

FACILITIES FOR CULTIVATING MILKY MUSHROOMS

CONTENTS

P.K. Khanna and S. Kapoor (2007) Manual on mushroom production. Dept. of Microbiology College of basic sciences and Humanities, PAU , Ludhiana PP. 58-62

Fig. 13. Fully grown calocybe indica

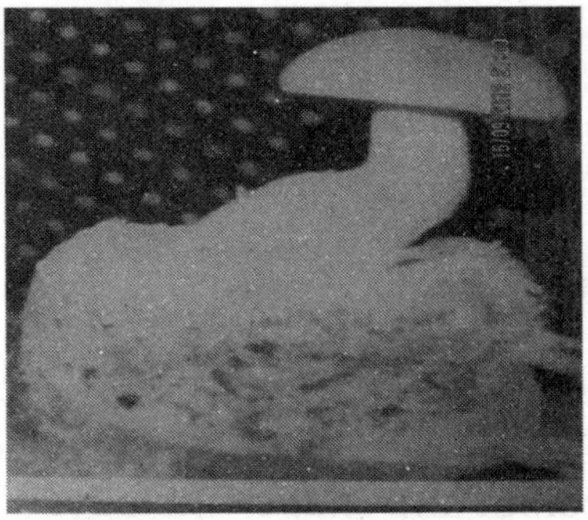

Fig. 14. Different stages of growing calocybe indica

Introduction

The milky mushroom technically known as *Calocybe indica* was first collected wild from West Bengal (India) by Purkayastha and Chandra in 1974. Later this mushroom was successfully domesticated and being grown as a cultivated variety in south India (Fig. 13 & 14]. The mushroom is known by diverse names in different parts of India and most common name *"Dudh Chhata"* implying milk umbrella due to its colour and shape. This mushroom can be cultivated in the temperature range of 25-38°C with good biological efficiency (60-70%) and has good potential under tropical and sub-tropical conditions. Hence it will fit well as a fill crop in the relay cropping followed under seasonal cultivation of mushrooms under Punjab conditions. In 2004 Punjab Agricultural University has recommended the cultivation of this mushroom variety during summer months from April to October.

Salient features of milky mushroom

Suitable for cultivation during summer months (April to October) throughout the state of Punjab and plains of North India (Temperature 28 - 38°C).

High Biological efficiency (BE) of about 50-60%, much higher than the only other tropical variety cultivated during summer i.e. *Volvariella volvacea* (paddy straw mushroom) which has a BE of only 12-15%.

Attractive milky white colour, large size fruit bodies and better shelf life (3-4 days at room temperature).

Good nutritive value (17% protein, 4% fats and essential minerals and vitamins) comparable to other winter varieties and summer varieties recommended by PAU.

Taxonomic classification

Kingdom	Fungi
Division	Basidiomycotina
Class	Basidiomycetes
Order	Agaricales
Family	Tricholomataceae
Genus	*Calocybe*
Species	*indica*

Biology

Morphology

This mushroom is commonly called "Dudh Chhata" implying milk umbrella due to its colour and shape. The pileus (cap) of the mushroom is white, 10.0–14.0 cm wide and non-hydrophanous in nature. At initial stage, cap is convex and later on it expands and becomes flat like an umbrella, stipe of the mushroom is in the centre but sometimes eccentric, cylindrical with subbulbous base. Generally, length of the stipe is about 10 cm and it's width vary from top to base. It is 1.8 cm thick at the top, 3.5 cm in the middle and 2.4 cm thick at the base. Hymenophoral trama, regular, with slight divergency below the sub-hymenium. Basidia tetrasterigmatic, clavate with carminophyllous granulation, 25.5 – 30.6 × 6.8 – 8.5 μm. Basidiospores ellipsoid showing white spore print.

Cultivation of Milky Mushroom

Requirements

The cultivation of milky mushroom has very simple requirements based on easily available raw materials namely wheat straw, Farm Yard Manure (FYM), polybags, spawn, covered growing space, drum for boiling.

Cultivation steps

1. Preparation of straw

The wheat straw is spread on a *pucca* floor and wetted for 16-20 h to attain 70-75% moisture level. The straw is then filled in a gunny bag and the bag is dipped in boiling water for 45-50 mins. During this boiling care is to be taken that the gunny bag should properly dip in the boiling water. The gunny bag is then taken out and the boiled straw is allowed to cool overnight and also to allow the excess of the water to drain out.

2. Filling of bags and spawning

The cooled straw is filled in polyethylene bags (12" ×18") @ 2kg wet substrate per bag. Multi-layered spawning of the wheat straw is done for spawning in case of milky mushroom. In this the container is filled up to 1–1.5" level with the wetted straw and bits of spawn are broadcasted on it and thereafter, gentle pressing is resorted to. In this way, the bag is filled with alternating layers of about 1-2" of wetted straw and spawned. Wheat grain based spawn is used @ 10% of dry weight of straw and each layer of straw should be spawned with approximately equal amount of spawn. After spawning, the upper end of the bag is tied with gunny thread and lower edges of the bag are cut to drain off the collected water. These bags are arranged in tiers on shelves. The spawn impregnates into the boiled straw in about 15-17 days.

3. Casing of the bags

After completion of spawn run, the mouth of the bags is opened and disinfected casing soil (disinfected as given for button mushroom) consisting of FYM (1 year old) : sandy soil (4:1; v/v) is put on the surface. The thickness of the casing soil should be approximately 1" and should not be too thick. The bags are kept on the shelves in the growing room and surface of the bags are watered daily twice a day. The temperature in the growing room should be between 25-38°C and humidity in excess of 75-80% should be maintained. The bags show appearance of pinheads in about 13-15 days.

4. Cropping

The pinheads or primordia develop into mature sporocarps or fruit bodies ready for harvest in tha next 5 days or so. At the time of harvest the cap diameter is about 5-7 cm and the weight of the fruit body is about 40 grams. If the fruit body is not harvested it can attain very big size and weight, however such large sized fruit bodies are not preferred in the market. The mushrooms are harvested by gentle twisting and lower part of the stipe with casing soil sticking to it is cut off. Mushrooms are marketed fresh in 200 g pouches. The cropping period continues for 35-40 days after first harvest. A yield of 50 to 60 kg fresh mushrooms can be obtained for 100 kg of dry straw used for this mushroom.

Wet whet straw overnight
↓
Boil wetted straw (45–50 mins) in a gunny bag
↓
Remove straw and cool for a few hours
↓
Spawn (multi-layered)@ 10% dry wt. of straw
↓
Fill in polybags (12" × 18")@ 2 kg wet straw/bag and close the bag
↓
Incubate in the growing room (15–18 days)
↓
Case the bags with FYM _ sandy loam soil (4:1. v/v)–1" thick
↓
Maintain high humidity in the room (75–80%)
↓
Spray water regularly on the top of the oosing layer
↓
Primordia initiation in 13–15 days
↓
First Harvest 5–6 days following pinning
↓
Cropping for 35 –40 days

FLOW CHART-CULIVATION TECHNOLOGY FOR MILKY MUSHROOM

Cropping schedule

The schedule of three crops may be followed as outlined below:

Activity	First crop	Second crop	Third crop
Wetting & Boiling of wheat straw	Starting April	*Mid June*	Mid August
Filling and spawning	Starting April	*Mid June*	Mid August
Casing of bags	Mid April	End June	Start September
Cropping	Start May - Mid June	Start July Mid August	Mid September End of season
Removal of spent	Mid June	Mid August	End of season

❑❑❑❑❑

CHAPTER **11**

PRESERVATION OF MUSHROOMS

CONTENTS

Mushrooms can be easily grown even by small farmers as their cultivation is economical and needs low cost technology, hygienic conditions, care and above all interest and enthusiasm. (Vijayakhader and Pandya, 1981 a). Like most of the agricultural products such as vegetables, fruits, and the like, mushrooms are perishable. One way to over come the perishability is to consume them fresh. This does not however mean that they cannot be preserved, when they need to be transported to long distances.

Harvested mushrooms keep fresh for 24 hours at room temperature. Pack them in polythene bags with a few holes here and there to keep them fresh. Keep the bags in a refrigerator at a temperature of 5° to 8°C or keep the bags in ice containers taking care to see that the shapes are not broken or spoiled, cold stored mushrooms will remain fresh for 4 to 5 days.Mushrooms can also be sun dried and preserved.

Mushrooms can be canned and pickled. Pickled mushrooms can be preserved for about an year, (Vijayakhader & Pandya, 1981 b).

Mushrooms can be preserved in brine or dried and transported in suitable packages.

Fresh mushrooms after picking are cleaned properly and then sliced longitudinally, dehydrated under hot air continuously for 4 to 5 hours at a temperature of 50° to 55°C. The dried mushrooms are crisp and are packed in airtight containers immediately to avoid re-absorption of moisuture.

Dried mushroom segments especially the small pieces can also be powdered, mixed with powdered pepper and cloves and used in tasty soups and sauces.

Sterilized mushrooms are preserved after heating on a low fire in their own juice or in tomato puree, in edible fat, in wine and or pickled in salt or vinegar.

Fig. 15. Dried Oyster Mushrooms

Another possible way is to prepare mushroom extract. Mushroom extract is a thickened juice obtained from fresh mushrooms by steaming. Usually well grown large sized mushrooms are more suitable for making an extract as they often contain a large amount of water. After mushrooms have been cleaned, sliced and washed, salt is added and when they are cooked on a low fire produce juice. When they start producing juice, this is poured out and thickened by further boiling. This extract is bottled and used after seasoning with pepper and cloves to taste.

Mushrooms can be preserved by pickling them in salt which is considered the most profitable method of retaining the unique flavour. Cleaned and well drained mushrooms are cut into small pieces, arranged into layers alternating the layers with salt These layers can be arranged in wide mouthed sterilised bottles, and packed air tight. Mushrooms preserved in this way have a longer shelf life. Mushrooms can be used in a variety of dishes.

The most popular method of preserving mushrooms is pickling them in vinegar. Cleaned and washed mushrooms are cut into pieces of desired size and are placed in a pan half filled with water. After adding salt, whole pepper corns and bay leaves, this mixture is then boiled on low flame for 20 minutes. They are then strained in a clean muslin cloth, mixed with the required quantity of sliced onions and the entire mixture is packed and preserved in air tight sterilized glass jars. Half boiled carrots can also be added to improve taste before bottling them, if desired.

Dehydration

The fresh mushrooms are thoroughly washed in cold running water until they are clean. The washed mushrooms are blanched in boiling water for 3-4 min by tying in a muslin cloth or putting in wiremesh basket and cooled in cold running water. The blanched mushroms are soaked in solution containing potassium metabisulphite and citric acid (3000 ppm each) for 15 min. The extra water is drained off by spreading the mushrooms on wiremesh trays. The mushroms are transferred to the trays of through-flow dried and spread in a thin layer (of 3-4 cm thickness) and dried at 55-65°C for 5 hr. and later on at 45-55°C for another 1 and a half to 2 hr. The dried mushrooms can be stored in an airtight container at room temperature [Fig. 15].

Dehydration ratio 13-14 : 1

Reconstitution ratio 1 : 5-6

Home Drier

A small drier is ideal for drying mushrooms at home. It consists of a strong box of galvanized iron sheet (90 cm. × 60 cm. × 90 cm) with perforated bottom. The sides and top of this box are enclosed in wooden frame work. The box is supported on charcoal hearth, or electric heater can be used as source of heat. Moisture is let off through two slits 60 cm. × 4 cm. with collapsible metallic flaps along the length on both sides about 10 cm. below the top. The drier can accommodate 7 trays of 80 cm × 60 cm. Care is taken to regulate the temperature and outflow of moisture from the drier by adjusting the source of heat and the slits (Fig 16.).

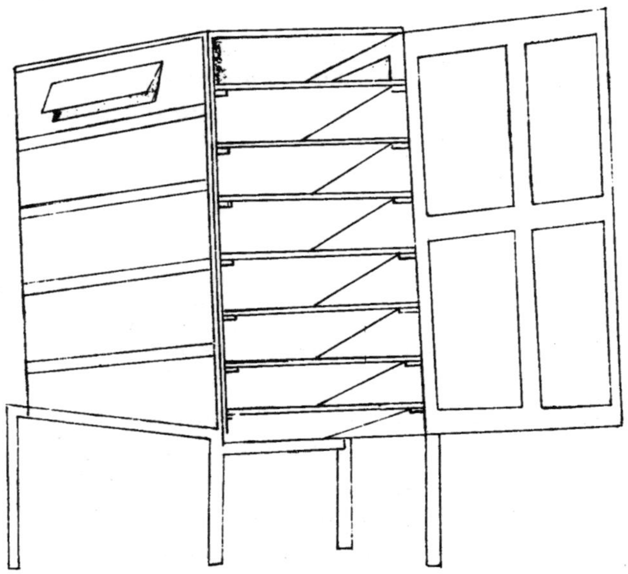

Fig. 15. Home Drier (CFTRI, 1990) for drying mushrooms

Dried mushrooms should be put into tins and sealed air-tight with solder or wax, depending upon the duration of storage.

Cabinet driers or truck driers can be used on large scale drying.

Freeze drying

A commonly adopted practice of commerce, where mushroon (button mushroom) are sliced, immersed in 0.05 per cent sodium metabisulphate and 2 per cent common salt soultion for 30 minutes. They are then blanched in boiling water for one or two minutes followed by cooling. This is then frozen for one or two minutes followed by cooling. This is then frozen for one minute at 12°C. The frozen mushrooms are dried to a moisture content of 3 per cent in freeze drier, and packed under vaccum.

Canning

The freshly harvested mushrooms are trimmed, thoroughly washed and spread on wire mesh trays having steel mesh, in a layer of 5-6 cm thickness. They are steam blanched by passing the free steam at 90°C for 1 min. and cooled at room temperature. The blanched mushrooms are graded according to their size or cut uniformly, filled in 8 oz plain cans (130 g in each can) filled with brine solution containing 2% sodium chloride and 0.1% citric acid, The cans are exhausted in an exhaust box and sealed. The sealed cans are processed at 15 lb pressure for 15 min. cooled in cold running water, wiped and kept in a dry place.

One kg of fresh mushroom gives 5 cans of 8 oz capacity.

Usual canning equipments, driers (through-flow) etc., which are available in any scientific stores.

The process can be released from CFTRI (1989), Mysore under consultancy basis on mutually agreed terms and conditions.

Canning is adopted on a very large scale for the preservation of button mushrooms. The following is the technique described by Sethi and Anand (1978).

1. Mushroom for canning are harvested, usually at an early stage. The stalks are cut and the selected buttons are to be of uniform size.
2. Wash the mushrooms properly to remove dirt and other foreign matter.
3. Blanching is necessary to preserve the colour for which the washed mushrooms are dipped in boiling water for 2-3 minutes, drained immediately and dipped in cool water.
4. Blanched and cooled mushrooms are filled in cans to about 3/4 capacity. Approximately about 200 g mushrooms are filled in standard 1 lb cans.
5 Salt soultion consisting of 2 per cent common salt, 2 per cent sugar and 0.3 per cent citric acid is boiled, filetered through muslin cloth and added to the can to just fill it up to the brim (125 ml solution for a lb can).
6 After placing the lid on the can they are kept in boiling water or steam till the temperature in the centre of the cans reach 80-85°C.
7. After placing the lids in position, they are sealed on a Seamer to get an air-tight

CHAPTER 12

FOOD VALUE OF MUSHROOMS

P.K. Khanna and S. Kapoor (2007) Manual on mushroom production. Dept. of Microbilogy college of basic sciences and Humanities, PAU , Ludhiana PP 82-83

Introduction

Cultivated mushrooms contain varied amount of proteins, carbohydrates, minerals and vitamins, and rank nutritionally between low grade vegetables and high grade meats. Although adequate nutrition is furnished by animal products, such as fish, meat, eggs and milk, in India, these articles are not within the means of the common man. Consequently, most of our population lives on grains, tubers and small quantity of pulses which supply proteins deficient in essential amino acids. In fact, the situation is more serious in our country because carbohydrates and fats, the main energy giving constituents of food, are also not easily available. Food crops in India suffer from natural calamities, but mushrooms, which are grown indoors, are not prone to these calamities (floods, drought, etc.). With the well-established techniques for mushroom growing, a predictable crop is possible and one can easily supplement one's daily diet by growing a few trays of mushrooms in a room. Thus, mushrooms which are rich in proteins, carbohydrates, minerals and vitamins have become accessible to the common man and can be appropriately referred to as mycomeat. Anyhow, mushrooms cannot be a substitute for meat, fish and eggs as far as their nutritional value is concerned. However, the fact remain that mushroom contain about 4% protein, but fish, meat and eggs contain 14.7-19.5%, 21.4% and 13.3% proteins, respectively. The protein content of mushrooms can be compared with that of vegetables and fruits. It has been found that the protein content of mushrooms is twice that of vegetables, such as asparagus and cabbage and 4 times and 12 times that of fruits, such as oranges and apples, respectively.

The ash analysis for the presence of minerals shows the presence of calcium, 0.0024%; phosphorus, 0.15%; potassium, 0.50%; total iron 19.50 ppm; available iron 5.95 ppm and copper, 1.35 ppm. Regarding the amino-acid composition, it contains 7 essential amino-acids, in addition to 11 others. In addition, mushrooms rank very high in respect of their vitamin content. They are known to be excellent source of vitamins thiamine (B_1), ribofiavin (B_2), niacin, pantothenic acid, biotin and folic acid.

Proximate composition of mushrooms (Dry wt. basis)

Proteins	25 – 37%
Carbohydrates	48 – 57%
Fats	1.0–3.2%
Crude fibre	11–14%
Ash	5 – 9%
Vitamins	B_1 & B_2
Minerals	K, Ca, P, Fe
Calories	261 – 385

Therefore, mushrooms are a good source of protein and supply varying amounts of carbohydrates, fats, minerals and vitamins to the human diet. Mushroom supplementation in the food also satisfies the energy requirements to some extent. Besides, contributing to vegetarian diet, mushrooms have additional medicinal benefits which are listed below:

1. Mushrooms are low calorie-food, their low carbohydrate and fat contents also make them ideal food for fat persons who wish to reduce their excessive weight.
2. Ascorbic acid is useful in combating scurvy and gingivitis in infants.
3. Niacin and pantothenic acid (water-soluble vitamins) are beneficial in pellagra, dermatitis and the burning of hands and feet.

❑❑❑❑❑

CHAPTER 13

MEDICINAL VALUE OF MUSHROOMS

P.K. Khanna and S. Kapoor (2007) Manual on mushroom production. Dept. of Microbilogy college of basic sciences and Humanities, Punjab Agricultural University, PP 5

Mushrooms have been used for medicinal and tonic purposes in China and Japan since ages. In recent years there has been a tremendous interest in the medicinal properties of mushrooms. Cosmetic products and beverages have been produced in China from *Ganoderma lucidum*. Pharmaceuticals worth millions are produced annually in Japan from *Lentinus, Coriolus, Schizophyllum* and *Ganoderma*. Some important medicinal mushrooms are given below:

Mushroom	Medicinal Effect
Amanita muscaria	Cure malfunction of GABA system
Armillaria mellea	Purgative
	Stomach & intestine recuperating
Coprinus comatus	Anti-diabetic
Ganoderma spp	Sedative
	Hepatoprotective
	Improve intelligence
Grifola frondosa	Beneficial in blood pressure, diabetes and Constipation

Mushroom	Medicinal Effect
Lentinus edodes	Liver protective
	Enhance vigour, energy, sexuality & diminishing age
Marasmus androsaceus	Analgesic & sedative
Phanas alombe	Stops gum salivation
	Good styptic
Pleurotus sajor-caju	Reduce rate of nephron deterioration
Psyclocybe mexicana	Mental disorder
Tremella fuciformis	Tonication
	Moisturize lungs
	Antiradiation drug
	Promote saliva secretion
Tricholoma populinum	Against hay fever
	Allergic blood vessel & inflammation
Volvariella volvacea	Hemolytic agent

❑❑❑❑❑

CHAPTER **14**

DISTINCTION BETWEEN EDIBLE AND POISONOUS MUSHROOMS

CONTENTS

- *Important edible mushrooms*
- *Important poisonous mushrooms*
- *Types of mushroom poisoning*
- *Tentative criteria to distinguish between edible and posionous mushroom*

P.K. Khanna (2003) Mushroom cultivation, Punjab Agricultural University, Ludhiana Chapter VI, PP 26-34

Mushrooms, as understood, include the fruiting bodies of all fairly large, higheer fleshy fungi, both edible and poisonous. However, originally, the word mushroom was used to refer to the edible ones, whereas the word toadstool was used for the poisonous ones of the gill 'fungi'.

For the easy placement of different mushrooms into their representative families, the following simple key may be helpful : Fructifications, cap like to shelf-like, having gills or lamellae, pores or teeth on the lower surface of cap (pileus).

Cap or pileus having tubes on the undersurface.	=	Pore mushrooms
Cap or pileus having gills or lamellae on the undersurface.	=	Gill mushrooms

Cap or pileus with teeth or spines	=	Tooth mushrooms
Fructifications not as above, but coral-like, ball-like or otherwise,		
Fructifications; coral or club-like	=	Coral mushrooms
Fructifications; ball-like, sometimes broken by the lengthening stalk, spore mass with or without spore	=	Puff ball and stinkhom
Fructifications; saddle-like morchelloid or or globular and mushroom subterranean	=	Morels and truffle

Important edible mushrooms

Over 2,000 species of fungi are reported to be edible. Of these, many have been collected, recorded and safely consumed in India. Some of the university accepted edible mushrooms are discussed below.

All species of the *Morchella* (morel's Guchi) are a highly prized table delicacy. They can be easily recognised by their large size; prominently stipitate, conical to globose pileus which, in turn, is pitted with large shallow to deep-grooved lines with hymenium. The pileus is invariably cream to yellowish to brown. The stipe is light-coloured, cylindrical, hollow and almost smooth variety. Five species namely, *M. angusticeps, M. esculenta, M. conica, M. hybrida* and *M. deliciosa* are known from India. Of these, *M. conica* and *M. esculenta* are the commonest. These species can be collected in the north-western Himalayas in spring (April and May).

Two species of *Helvetia* namely *H. Crispa* and *H, elastica are* edible. They can be easily recognised by their saddle-shaped, usually fragile pileus mounted on a permanent stipe. In *H. crispa,* the pileus is white to pale yellow, the stipe is long with prominent branched ridges. It is one of the commonest species, growing luxuriantly on moist soil in the Himalayas. In the case of *H. eslastica,* the pileus, or cap, is brown to black and the stipe is very long, cylindrical and without ridges. Both the species are easily collected during the monsoon months, viz. July-September, in the north-western Himalayas.

Another group of edible mushrooms included in the Discomycetes are tuber-shaped, dark, sessile fruiting bodies, which are commonly referred to as truffles. These mushrooms are usually collected by employing trained dogs or pigs, that recognize their characteristic smell. Tubers are found in the soil among the roots of oak-trees. They can also be detected because of the cracking of the soil above them or because of the poor and scorched look of the overgrowing vegetation The two most sought after truffles are *Tubur melanosporum* (Perigord truffle) and *T. magnatum* (white tuffle) found in France and Italy, respectively.

Some of the hydnums (tooth fungi) are among the best edible fungi and are commonly known as hedge hog mushrooms. The members of *Hericium,* namely *H. coralloides* and *H. erinaceus are* eaten in different parts of the world. Both the species are abundant in the north-eastern Himalayas and can be conveniently collected in monsoon months, viz. July-September. *Hericium* can be easily recognized by their nodular and cushion like or much branched fleshiy fruiting bodies with pendant spines. The true stipe may not be present. *Hericum erinaceus* is easily distinguished from *H. coralloides* by its cushion-like sparingly branched fruiting bodies and very long spines.

Hydnum repandum is another edible fungus, which is very common in the north-western Himalayas. It can be easily recognised by its light cream to creamy-buff to deep buff fruiting bodies and non-decurrent spines. Along the coral fungi, all clavarias are safe to eat, except *Clacaria dichotima. Sparassis crisspa* is another important edible mushroom which is popularly known as "wood goat" in the hills. It is white to creamy yellow, with numerous thin and somewhat flattened branches. It is found in the vicinity of oak-trees. Boletus edulis, one of the tube fungi, is safe to eat. Its white flesh does not turn blue and the stem is covered with a fine white network. The tubes are creamy white in young fruit bodies, but become greenish later on. *B. pinicola* and B. cyanescens are also worth eating. Among the bracker or pore fungi, three or four species are tender to serve as food. These are *Polyporus sulphureus, P. frondosous* and P. *umbellatus.*

Among the Gasteromycetes, puff balls are safe to eat but great caution may be used to see that their interior is pure white. Many species of *Calvatia* (*e.g. C. gigantea*) produce big fruiting bodies which are often big enough for a large family to consume. It may be up to 500 cm in diameter and is the largest fruiting body in the puff-ball family. *Lycoperdon* spp. are also edible in the young stage but the spores start ripening soon. Consequently, it is difficult to find them at the right stage. *Lycoperdon fusilum,* with aromatic odour and pleasant taste, is liked by everybody. This species is found in abundance in the Punjab during July-September. *Podaxis pistillar is* commonly known as *"khumb"* is also found in abundance during the monsoon months. It has a'slender woody stipe, arising from a narrow volva and bears a more or less pyriform or rounded head. The center of the stipe penetrates into the glebate apical portion of the peridium (the covering of the gleba). At maturity, the peridum pulls from the stipe below and splits near the latter, exposing the powdery glebs.

Among the white spored "gill fungi". *Lepiota procera,* the porosal mushroom is good for eating. However, it resembles the deadly *Amanita* in some respects. It is characterised by a reddish-brown, flat umbonate cap beset with prominent, concentrically arranged scales, except on the unbo. The stem, tapering from a bulbous base, is usually hollow, with prominent and somewhat movable ring without a volva.

All species of *Pleurotus* are lignicolous and grow in clusters under natural conditions. The stipe whenever present, is lateral, rarely central. Two species mostly sought after for edible purposes, are *Pleurotus ostreatus* (oyster mushroom) with a white spore print and P. sapidus with a liliac spore print.

In the Punjab Agricultural University, Ludhiana, however, the simple cultivation technology for raising P. *florida, P. sajor- caju* and *P. ostreatus* on chopped unfermented wheat straw (*turi*/paddy straw under Punjab conditions) has been developed.

Two *Lactarius* spp. namely L *deliciosus* and *L. volemus* are milk bearing agaries and edible. The former with saffron- coloured milk, is common in distribution. *L. volemus* is characterised by profuse white milk and white or pale brown gills, becoming darker with age or when bruised.

All species of *Cantharellus* are edible. The genus is characterised by a funnel-shaped cap and decurrently formed gills and white to creamy spores. It is distinguished from *Clitocybe* \ by anastomosing dull-edged gills. *Clitocybe cibriarius,* the golden chantrelle, has been a favourite delicacy. It is a small mushroom, with a trumpet like shape, a solid body with a blunt edge, much interwoven gills, with colour like that of the yolk of an egg fried hard. Members of *Coprinus* serve as a good food and are the most easily recognisables, since the substance of the gills deliquesces and dissolves by a process, known as autodigestion. Various species can be recognised by their inky cap (fruiting bodies). *Coprinus comatus,* popularly known as . shaggy man, is very common and has been reported from the Punjab State.

Among the purple-brown spored agarics *Psalliota* spp. are edible and are among the most prized dishes in the world. *Psalliota campestris* (*A. bisporus/A.brunescenes*) grows in meadows and is short and squantty. It possesses only a slight ring and there is no volva. This mushroom is commonly called the temperate mushroom, the white button mushroom, the European mushroom and so on. It is cultivated on a commercial scale throughout, the world.

Dr. E F K Mantel, an FAO Expert, has been a great force in initiating mushroom growing in India. Successful cultivation of mushrooms in India is restricted to a few genera of the gill fungi. Mention may be made primarily of *Agaricus, Volvariella* and *Pleurotus.*

Important poisonous mushrooms

There are very few poisonous mushrooms, even if we use the term "poisonous" in its widest sense., Most of the fatal cases assigned to mushroom eating are due to *Amanita phalloides* (the death cap) or its ally (the destroying angel), *A. pantherina* (the Panther mushroom) and *A. muscaria* (the fly agaric) as well as *Galerina* spp.

The genus *Amanita* is characterised by the presence of both the ring and the volva.

A simple key for the separation of the 5 species *of Amanita* is given below :

1.	Fruiting bodies, white throughout	2
2.	Fruiting bodies, not white throughout	3
3.	Cap, with an umbonate center, fruiting bodies fairly large in size	*A. virosa*
4.	Cap, with an umbonate center, fruiting bodies fairly small	A *verna*
5.	Cap, smooth, margin even and not striate	*A. phalloides*
6.	Cap, beset with prominen⁺ warts, margin striate	4

7.	Stem, smooth, volva evanescent cap not brightly coloured	*A. pantherina*
8.	Stem prominently sqamulose in the lower part, volva broken into scales of distinct rings in the lower part, cap brightly coloured.	*A. muscaria*

Most of the deaths occur owing to the consumption of *A. phalloides* and 39 deaths in England between 1920 and 1950 are known to have occurred owing to the eating of this species. In the USA', on the average, about 50 fatal cases reported every year, are assigned to the eating of death cap. The symptoms of poisoning due to *A. phalloides* are characteistics and terrible. A ten to twelve hours of incubation period is followed by sudden abdominal pains, vomitting, cold, sweat, diarrhoea and excessive thirst. The'symptoms subside after 2 days and occur again. After the paralysis of the nervous systyem and degeneration of the liver, death occurs in 3 to 10 days after ingetion. All parts of the fruiting body are dangerous, including the spores. Recovery is difficult. Ninety per cent of recorded deaths are due to *A. phalloides.*

Amanita muscaria or fly agaric is also known to cause deaths even among healthy persons. After ingestion, the incubation period is 2 to 3 hours. Illness manifests itself as colic pains, vomiting and diarrhoea, loss of memory and tendency to sleep. The poison primarily resides in the skin of the cap which peel off. It was formerly used in certain countries or getting rid of bugs and flies. The active toxic principle is the mycetomuscarine and muscardine. The dried powder ortincture made from the flesh of the fungus is used in homeopathic for curing sunstroke. *A. panthrerina* resembles, *A. muscaria* in its effects and chemistry.

Inocybe patouidare

It is a common mushroom of the deciduous woods. It resembles the field mushroom in its younger stage, but the cap is conical, with an enrolled margin and there is no ring on the stipe. It immediately strains during handling and with age changes to brownish red. Other species of the genus also redden on touching, but are difficult to distinguish without microscopic examination. Symptoms are like those of muscarine poisoning.

Entoloma lividum

This species is known to have caused many accidents in France. However, death rarely occurs afrer consumption, at least never in the case of healthy persons. It can be distinguished from the field mushrooms by the absence of the ring on the stipe, it causes violent sickness and diarrhoea as characteristic symptoms in human beings.

Entolomo sinuatum

This species is also poisonous and produces headache, stomach pains and vomiting. The two species *E. lividum* and E. *sinuatum* are differentiated on the basis of the hollow stem present in the former.

Gyromitra esculenta

This specie is characterized by a chestnut red cap, with brain like convolutions, and with a whitish stuffed or hollow stipe, and binucleate, large yellowish spores. There are controversial reports about its poisonous nature. There are persons who have consumed it without all ill effects. Its poisonous principle is helvellic acid, which is soluble in hot water; hence cooked and dried samples are safe.

Types of mushroom poisoning

Dickinson and Lucas (1979) have recorded 5 different types of mushroom poisoning:

1. The cellular poisons : *Amanha* species and several small *Galerina* species attack tissues of the body and are responsible for many deaths. Besides these species, the false morel (*Gyromitra esculenta*) also falls in this category and contains Gyromitrin- a toxin which breaks down the red blood cells but this toxin can be removed before cooking.

2. The gastric poisons : Several mushrooms fall under this category and the early symptoms are occurrence of nausea, vomitting and diarrhoea which terminate rapidly and the patient recovers in a day or two from poisoning due to *Boletus satanas, Entoloma lividum, Lactarius torminosus* and *Russula* spp. However, toxic substance in these mushrooms can be removed by cooking.

3. The nerve poisons : Gasterointestinal symptoms appear in 2-3 hours after consumption resulting in violent convulsions coma often death, e.g. *Amanita muacaria, A: panthehna Inocybe patouildare, Clitocybe* spp. Muscarine and ibotenic acid toxins contained in the poisonous mushroom are responsible for the above symptoms.

4. Mushrooms and alcohol: *Coprinus atramentarius* (a wild mushroom) the common ink cap if eaten alone has no ill effect. However, if alcohol is taken, severe sporadic reaction sets in and the symptoms are like the one induced by diesulfiram (anatbuse), a synthetic chemical which is used for treating chronic alcoholics. However, shaggy ink cap (*Coprinus comatus*) does not have any ill effects.

5. Hallucinogenic mushrooms : *Amanita muscaria* the fly agaric, has been used as a ritual haliucinogenic in certain cultures of the world. This mushroom contains ibotenic acid and musciol as principal toxin responsible for sensory distortions leading to hallucinations.

Tentative criteria to distinguish between edible and posionous mushroom

For many years, people have used mushrooms as food and from time to time deaths have occurred from their use, either through carelessness or through ignorance. Those who use them want to know how to distinguish between poisonous and non-poisonous ones. Though many attempts nave been made to distinguish edible mushrooms from poisonous toadstools, the only foolproof method to do so is to learn to recognise each sample by its specific characters. In case of doubts about the identity of a particular edible species, it is better not to use them untill all doubts about the identity of a particular edible species, are cleared by a specialist, or by a person who is familiar with them. No effort should be made to try a new species because this may prove to be expensive experiment, resulting in dire consequences. However, for the beginners, some criteria can be used for separating edible and poisonous mushrooms.

1. While collecting edible mushrooms, immature fruiting bodies may be avoided, because in the young stage, they do not clearly show all the characteristics of the species to which they belong.
2. The mushrooms with a cap or sac like envelope (volva) should not be consumed. However, *Amanita caesaria, A. rubercens* despite with a sac like envelope at the base of the stem are edible.
3. Those with a milky juice are not fit for eating, but *Lactarics volemus,* with abandant white milk, is edible.
4. In the case of the pore fungi (*Boletaceae*), avoid those species whose flesh changes colour when bruised or cut. However, *Sporobolomyces strobilaceus* changes colour, when cut, but is edible.
5. Avoid over-ripe mushrooms or those in the process off decay.
6. All puff balls are edible, when young.
7. Try to familiarize yourself thoroughly, with a few species at first and then gradually extend your knowledge to others.
8. Do not make any attempt to eat a large number of different" species in one season.

It can therefore, be concluded that none of the criteria to differentiate between good and bad mushrooms is satisfactory and the best course to learn to recognise each species is with the help of standard monographs and to get the samples identified by a specialist, at least once, in order to be sure.

CHAPTER **15**

DISEASES OF MUSHROOMS

CONTENTS

P.K. Khanna and S. Kapoor (2007) Manual on Mushroom Production. Dept. of Microbiology, College of basic sciences and Humanities, PAU , Ludhiana PP 64-73

Introduction

The diseases, in general, may be due to the fault in preparing compost, casing soil, bad sanitary conditions and variations in humidity and temperature prevailing in the mushroom houses.

Fungal Diseases

In general, the control of the fungal diseases of mushrooms is difficult. The short-lived nature of the mushroom crop also renders their control more difficult and once a disease (fungal, bacterial or viral) appears, there is hardly any time to control it. Consequently, stress is laid on the preventive measures against mushroom diseases rather than on curative ones. It is through correct cultural practices that many maladies of mushrooms can be prevented. The diseases may be pathogenic (fungal, bacterial and viral) or non-pathogenic (environmental, chemical, etc.). Some important diseases and competitors of button mushroom, which are of relevance in India are discussed here.

Dry bubble, brown spot, fungus spot or *Verticillium* disease:

This disease is caused by *Verticillium fungicola,* which produces light brown spots on the mushroom cap. These brown spots coalesce and cover the caps in an irregular brown blotch which sometimes is incorrectly diagnosed as bacterial blotch. These brown spots, on microscopic examination, show a fine white to greyish mould, having unicellular, hyaline conidia, borne in terminal clusters or branched verticillate condiophores.

In more severe cases, mushrooms (caps and stalks) become distorted. The infected stalks/stipes show splitting or peeling back but the lesions may not develop. In addition, infected fruiting bodies may be malformed into puff ball like shape without distinct stipe, commonly called bubble or dry bubble and the mould may cover the entire surface of the mushroom pins. Besides this, the affected mushrooms become leathery, dry and have offensive odour and are deformed.

Control

It is better to take up preventive measures before the appearance of the disease, because once it appears, it is difficult to control it. The primary source of infection appears to be the casing soil and inoculum (mycelium and spores) is carried by insects, air, water droplets, workers, etc. The inoculum builds up on 'trash' (portions of fruiting bodies left on the casing), involuntary dropped mushrooms on the floor of the growing rooms, on the tools used in management and on the infested casing soil, etc. As a preventive measure either carefully steam pasteurise or chemically disinfect the casing soil.

High temperature during cropping should be avoided and adequate ventilation is an effective check against this disease. Three sprays of Dithane M-45 (0.25-0.5%) give a good control of this disease. The first spray should be at casing, the second one at the time of prophylactic pin-head formation and the 3rd one after two flushes of the crop. Besides this the use of 15% Zineb dust can also prevent the disease. Sodium hypochlorite (15% dust) also helps in reducing the spread of the pathogen.

Wet bubbles, white mould or *Mycogone* disease:

This disease is caused by *Mycogone perniciosa* which produces an abundant white mouldy growth leading to putrefaction (giving foul odour) of the mushrooms with a golden brown liquid exudate containing bacteria and spores, the reason for the disease being commonly known as wet bubble. Wet bubbles attain a large size as compared to dry bubble. More severely attached mushrooms become distorted and are reduced to a more or less globular mass covered with mycelium, becoming necrotic followed by wet, soft rot giving an offensive odour. However, the swollen stalk and the unusually small cap also indicate the mycogone infection. If the infected mushroom tissue is examined microscopically, two kinds of spores can be seen:

Control

This fungus survives in soil and on other fungi besides being parasitic to wild mushrooms. Like dry bubble, its infection is carried primarily through casing soil but the introduction of the pathogen through other agencies, e.g., spent compost and infested trash is not ruled out. The infection can be carried by air, water, flies and workers, etc. Properly pasteurised, chemically disinfected casing soil is an important operation for avoiding this disease. Besides good hygienic practices, the use of Dithane M-45 (as in case of *Verticiilium* disease) also helps in combating mycogone disease.

Mildew, soft decay or cobweb disease:

This disease is caused by *Dactylium dendroides* and is also known as *Dactylium* mildew disease. The pathogen develops into conspicuous cottony growth spreading over casing as well as on the mushrooms within 12 hours. Large patches (8-18" in diameter) of mildew appear on the beds and all buttons in this area are infected and become unsalable. The pathogen possesses branched, septate, grey to white mycelium which develops to a rosy-red or yellow blue colouration. The conidia are yellowish, two or multi-celled and are borne on branched conidiophores. Infected mushrooms develop a soft wet rot internally.

Control

The pathogen harbours on wild mushrooms, field soil, spent compost and improperly disinfected casing soil. The treatments already mentioned for *Verticillium* and *Mycogone* diseases are also effective in this case. In addition, the mushroom house must be disinfected with 5% formalin solution before putting trays for the next season. Good ventilation should be provided and high humidity in the mushroom house should be avoided to check this disease.

Trichoderma spot, Trichoderma blotch, Trichoderma mildew/green mould:

The Trichoderma diseases are caused by *T. viride* and *T. koningii* which can be isolated from compost and casing. The former secretes a toxin into the casing, killing the mushroom tissues resulting in reddish brown to mahogany brown discoloration along the stipe. In addition, this mould is characterised by its dark green colour and by causing dry sunken non-coalescing lesions on the mushroom caps. It can appear at any time during cropping. *Trichoderma koningii* appears at late stage of cropping and the buttons overgrown by mould show soft wet rot. This mould is characterised by white downy mycelium with clusters of sage green spores. *Trichoderma* spp. harbour in composts as weed moulds or indicator moulds with inadequate nitrogen supplementation. From compost the *Trichoderma* spp. penetrate casing and produce large circular colonies on its surface, leading to considerable delay in the crop production.

Control

These moulds harbour in compost, casing and untreated wood surface etc. For effective control, the compost and casing should be carefully prepared besides using dithiocarbamate and benzimidazole fungicides.

Truffle:

It has been recorded to be one of the commonest competitors of the white-button mushroom but the fungus (*Diehliomyces \ microsporus*) is not the edible truffle. Its mycelium is white to yellowish, scanty at first, but later organizing itself into whitish to cream, solid, convoluted and wrinkled masses resembling calf's brain. These are the fruiting bodies of the truffle, ranging from 1mm to 3.5cm in size. The fruiting bodies contain oval to sub spherical asci, each with 8 smooth, subglobose, yellowish (sulphur-coloured) spores.

Control

This fungus, which is antagonistic to the growth of white button mushroom, can be kept under check through cultural practices. Consequently, adequate ventilation should be provided and high humidity in the mushroom house should be avoided to keep the truffle under control. The fungus inhibits the spawn growth and the compost becomes soggy and dark brown hence the production is greatly reduced. The best measure to combat this fungus is by altering the spawn running (to 18°C) and cropping (not more than 17°C) temperature in the mushroom house. Soil and straw smeared with soil during harvest, transport, storage and subsequently used for compost preparation are the major carriers of this fungus. Preparing compost from contaminant-free straw and avoiding contact with infested soil can help in keeping a check on the spread of truffle.

Plaster Moulds:

There are two fungi, namely, *Scopulariopsis fimicola* and *Populospora byssina,* which are competitors of the white button mushroom. Of these, the former is white in colour throughout; having branched, septate mycelium which bears one-celled conidia on phialides. Dense, white patches of the mould have been observed which first appear near the supports and corners of the trays but later on spread to other areas on the compost.

The brown plaster mould, caused by *Populaspora byssina,* first appears on the compost surface, starting at the supports and corners of the trays. Later on, large patches (6-15" diameter) are formed having dense white plaster like appearance which become yellow to brown or rust with age. The fungus is characterised by brown bulbils which are spherical in shape and are inter-woven with a fine network of white hyphae which gives waxy feeling on rubbing. These bulbils (sclerotia) are either air-borne or are carried through improperly prepared compost. Overwet compost and high temperature (28-32°C) during spawn should be avoided for effective control of the plaster moulds. If the infected patches of the moulds appear, even then, treatment with 2% formalin solution gives complete control of the disease.

However, mixing of 10 ppm (a.i.) of carbendazim (Bavistin/Derosal) with compost at spawning prevents the appearance of plaster mould.

Inky caps:

Inky caps include generally two species of *Coprinus* namely, *C. comatus* and *C. lagopus.* They occur as weeds of both the white button mushroom and the paddy straw mushroom and are characterised by long slender stipes and thin caps of the fruiting bodies. They frequently appear in the compost, on the spawned beds or even after casing in *A. bisporus* and in paddy beds laid for the cultivation of *V. diplasia /V. volvacea.* The pileus of the *Coprinus* spp. especially the lamellae, soon decay forming a blackish inky material (hence they are commonly known as inky caps) through autodeliquescence. This competitor can be kept under check by using properly prepared compost, which is free from ammonia. However, in addition, the grower must rogue the fruiting bodies of these weed mushrooms to prevent them from spreading further and reducing the yield of the cultivated mushroom.

Peziza spp:

This fungus is also a competitor and can be recognized by buff to dull-yellow fruiting bodies containing many spores in asci. The ascocarps of this weed appear on over- irrigated and improperly prepared compost used during the cropping days.

Besides the above mentioned diseases and competitors, there are several others known to hamper the mushroom industry in the world. However, as pointed out earlier, this industry being of recent introduction, not much work has been done on these diseases and competitors in India.

Bacterial Disease

Bacterial blotch, brown blotch, bacterial spot disease:

This disease is caused by *Pseudomonas fluorescens* biotype G, which forms pale yellow lesions on the caps at first, which become golden yellow or chocolate brown blemishes after enlarging and coalescing later on. These spots can be seen at, or near the edge of mushroom caps, at the contact points between them or wherever the caps remain wet for 4-6 hours after watering. However, during dry conditions, infected caps show radial cracking. The underlying mushroom tissue may appear as water-soaked and is grey in colour. Similarly mushroom stems also become blemished.

Control

The infection is initiated through the casing soil and other contaminated articles used in mushroom cultivation. It spreads through the splashing of water (during watering) or through flies, mites, workers during picking, etc. For the control of disease, pasteurize/ chemically disinfect the casing properly and provide good ventilation after watering so that caps of the mushrooms do not remain wet. The use of sodium hypochlorite (50ppm chlorine water) also reduces the incidence of this disease, provided the mushrooms are kept dry.

Viral Disease

Though none of the viral maladies of mushrooms are known in India so far, yet some important viral diseases recorded in other countries are mentioned for the general information of the grower. A number of disorders, the "La France complex", including "La France disease' brown disease, dieback, watery stipe, X-disease whose causes were unknown until recently, are most probably due to virus infection. The virus diseases manifest themselves in various forms depending upon the strain of mushroom cultivated, stage of its development, environmental conditions, intensity and virulence of the pathogen. However, the most consistent symptoms are:

(*i*) barren zones (no fruiting) on the beds;

(*ii*) degenerated growth in PDA of the mycelium from the diseased mushroom;

(*iii*) disappearance of mycelium from the casing soil and compost;

(*iv*) fruiting below casing soil;

(*v*) delayed appearance of pin heads of the first flush;

(*vi*) early maturity of sporophores and dwarfing, slow growth and very thick stipes (drum sticks);

(*vii*) rapidly dying mushrooms followed by wet, soft rot and

(*viii*) malformed gills or their absence.

Control

Viruses are mainty transmitted to healthy mushroom mycelium by infected mushroom spores as well as by infected mushroom tissue or mycelium. Since knowledge regarding the spread of the virus is very meagre, the rational control would depend upon more information about the natural vectors. Some control measures are given below:

(*i*) Heat the growing-room along with trays and compost at 70°C for 12 hours at the end of the crop.

(*ii*) Spray the trays and the other equipment with 4% sodium pentachlorophenate solution to which 0.5-1% soda (Na_2CO_3) has been added.

(*iii*) Disinfect the doors, floor, walls, etc. with 4% formaldehyde solution.

Non-pathogertio maladies

Rose Comb:

This abnormality is caused by frequent pesticidal sprays and by the incomplete burning of coal or kerosene used for heating the mushroom house. In this case, the cap is distorted and the gills or lamellae develop on its upper surface.

Long-stalked mushroom:

In a poorly ventilated mushroom-house, this type of abnormality is common. The stalk becomes very long whereas the cap remains small.

Scales:

In this non-pathogenic malady, the cap of the mushroom is not smooth but is covered with scales, which are due to low humidity in the growing room as well as excessive air currents in the growing room.

Stroma:

Aggregations of mushroom mycelium on the spawned compost or casing are known as stroma, which is basically related to the genetic character of the spawn. However, if the spawn is exposed to harmful petroleum based fumes or chemicals or detergents during its preparation, storage or transit also leads to stroma manifestation. It has been observed that. initial thin growth of the mycelium, in turn, changes into thick compact masses which hinder the formation of buttons. It may be mentioned here that spawn overgrowth on the casing due to abnormal environment or shallow casing should not be confused with the stroma. Patches of stroma should be removed and such portions on the trays may be recased, because usually no buttons develop on stroma.

❑❑❑❑❑

CHAPTER 16

PESTS OF MUSHROOMS

CONTENTS

P.K. Khanna and S. Kapoor (2007) Manual on mushroom production. Dept. of Microbilogy college of basic sciences and Humanities, PAU , Ludhiana PP 75-80

INSECT PESTS

When grown in modern mushroom houses, cultivated mushrooms are not affected by many pests. However, under seasonal cultivation as is carried out by many of the growers in India, mushrooms are highly susceptible to the attack of various pests. The major pests include:

(*i*) Mushroom flies - Phorids, Sciarids and Cecids

(*ii*) Mites (*iii*) Springtails

(*iv*) Beetles and moths

1. Phorid flies

Phorid flies are 2 to 3 mm in size when fully grown. They are brownish-black, hunchbacked and have short inconspicuous antenna and a reduced wing-spotting.

Phorid flies associated with mushrooms belong to the genus *Megaselia* and cause less damage to button mushroom as compared to sciarid flies. The most common and widespread species is *M. halterata, M. nigra,* and M. *agarici.* Other species include *M. bovista, M. longipennis, M, iroquoiana,* and *M. sandhui.*

Damage by Megaselia halterata

The female, after mating, is attracted by the volatiles from actively growing mycelium and it lays smooth, elongated ovoid eggs in the compost. Eggs hatch into larvae in 2-3 days which are nearly transparent and taper towards the head. The larvae feed gregariously on the mushroom mycelium and also tunnel through the stipe of the mushroom and move upwards towards the cap. During October and November, this fly attacks early crop of button mushroom but it becomes inactive during severe winter months. It again becomes active in March. The larval cycle is completed in about 14 days at 18°C to form the pupae, which are whitish initially but become dark yellow brown later. Adult flies emerge from the pupae in about 20 days at 18°C.

Control

In addition to hygienic measures such as sealing of cracks and crevices, proper peak heating and cooking out followed by proper disposal of spent compost and casing soil from previous crops can significantly reduce the number of phorid flies around a mushroom farm. Cook-out of the spent compost and casing material at 71 EC with steam for 2-3 h is effective in killing immature stages of the fly. The screening of doors and windows with mesh (14-16 mesh/cm^2) prevents the entry into the growing room. The use of different insecticides like diazinon, malathion, lindane, chloropyriphos and aldicarb incorporated into the compost act as a prophylactic control while dichlorvos is suggested for curative measures.

2. Sciarid flies

Sciarid flies are mosquito-type flies, which are 2 to 5 mm in size, they are black and have long upstanding antennae. The wing spotting is not so inconspicuous as in the case of phorid flies. Numerous species have been associated with cultivated mushrooms. The principle sciarid species are *Lycoriella auripila, L. mali, L. agarici* and *Bradysia tritici.*

Female sciarids lay upto 170 eggs in the compost or casing, but generally about 15 eggs are laid in one cluster. The development of eggs into larvae largely depends upon temperature. At 18°C, it takes about 6 days where as the time is much less at 24°C. Fully grown larvae (in about 18 days) are 5-8 mm long, white with a black shiny head. The larvae feeds on the mycelium in the compost and/or casing and also attacks young pinheads. Later they eat into the stems of the fruit bodies which turn brown and leathery making then unfit for sale. The pupae moves by convulsive movements rather than by typical worm-like movement of the larvae. The pupal stage is a resting stage and mature flies emerge in about 6 days at 18°C.

Control

For control of sciarid flies the proper disposal of spent material and screening of doors and windows as discussed for phorid flies is very effective. Poor quality of the compost tends to attract sciarid flies which lay their eggs in such compost. As a prophylactic control measure the use of Aldicarb, carbofuran, chloropyriphos, pirimiphos and lindane at dosages ranging from 12.5 to 100 ppm have been suggested. Chloropyriphos at 50 ppm dosage in compost and 25 ppm in casing soil has a good potential for fly control. The curative control measures include the use of various insecticides but dichlorvos has been found most effective for the knockdown effect with least persistence toxicity. In the mushroom house, dichlorvos in ultra low volume (ULV) spray applied with spinning disc sprayer was effective at 22.5 g a.i./100 m^3.

3. Cecid flies

Cecid flies, commonly known as gall midges, are dark brown and about 0.7-1.5 mm long and rarely seen in the adult stage. This is due to an unusual reproduction by "paedogenesis" (reproduction by immature forms). The fully grown fusiform larvae (2-3 mm long) produce daughter larvae so the generation time by this mode of reproduction at 24°C is only 6-7days. Each larva is capable of producing a many as 20 daughter larvae. Under other circumstances, the paedo-genetically produced larvae can switch over to sexual cycle to produce adult cecids via a pupal stage. Two species that are important are *Mycophila speyeri* which produces orange to orange-yellow larvae about 1 mm long and *Heteropeza pygmaea* producing white larvae about 3 mm long.

The larvae feed on the mycelium and can also eat vertical grooves in the stems. The white larvae of *H. pygmaea* crawls on the moist casing soil eating mycelium but causes limited damage to the mushrooms.

Control

General cleanliness, good pasteurization during Phase II of composting and effective disposal of spent compost and casing material or their cook out checks the infestation by this pest.

4. Mites

Mites are small spidery insects, which live on moulds, eelworms or bacteria in half-decayed or rotting organic matter. They are generally present in improperly prepared compost. Not all mites are pests of mushrooms. Only a few of them are mycophagous, other are saprophagous while still others are predatory mites which feed on mushroom pests and indicate their presence in the compost or mushroom beds. There are more than 50 mites species which are found associated with mushrooms in various parts of the world and four species have been studied in detail in India.

The important mites of mushrooms are

1. Straw or hay mites (*Tyrophagus putrescentiae*)– These are small, yellow-white mites,hardly visible to the naked eye. It bears long brownish-coloured hairs on the back and legs. They feed on moulds growing in compost. This mite feeds

on mushroom mycelium and mushroom tissue causing pits in the cap of the mushroom.

2. White mushroom mites (*Caloglyphus mycophagus*)– These are related to the straw mites, living in the same way and cannot be distinguished without the use of microscope.
3. Small mushroom mites (*Tarsonemus myceliophagus*)– These are tiny beige colouredmites, which are related to strawberry mites. They gnaw though the mycelial threads at the base of the stipe resulting in reddish brown colour of the base. It also feeds on weed moulds.
4. Pygmy mites (*Pygmephorus* spp.)– These mites are virtually harmless to mushrooms. They feed on weed moulds and are known as mycophagous mites.
5. Predatory mites are also called gamasid mites. They feed on the various stages of other mushroom pests. They are beneficial and are not pests of mushrooms.

Control

The control of mushroom mites can be achieved through efficient composting. Many fungal and bacterial contaminants present at the end of cropping in the spent material and saprophagous mites breed easily in such substrates. Cook-out of the spent material followed by sanitation of the mushroom growing rooms with Dicofol (0.1%) avoid the incidence of mites.

If mites are present in compost, a spray of diazinon (1.5-2.0 ml/10 litres of water) at filling is suggested.

Springtails

Springtails damage button mushroom by feeding on the mushroom mycelium in the compost. The attack on fruit bodies causes slight pitting and browning at the feeding sites. There is destruction of gill linings followed by development of a webby structure, scrapping of the stalk and withered primordia,

Control

The control measures for these springtails include, disinfection of composting yard with 0.05% malathion emulsion, incorporation of diaznon in compost at 30 ppm at filling and spray of malathion/dichlorovos at 0.025-0.05% concentration at spawn run or cropping period in case of infestation.

Beetles and moths

Beetles (*Staphylinus* sp, *Alphitobious laevigatus, Cyllodes wheteii*) and noctuid moth are the important beetles and moth affecting mushroom crop in India. The beetles caused small irregular holes in hymenium and stipe while the moth results in tunnels and pits in the mushrooms. Cleanliness and removal of spent debris away from the farm prevent the build up of population of beetles and moth.

Nematodes

Nematodes encountered in mushroom compost/beds are categorized as myceliophagus, saprophagous, predaceous, plant parasites and animal parasites. Among these, only myceliophagous and saprophagous are important as they affect crop yield through direct feeding or indirectly. *Ditylenchus myceliophagous* and *Aphelenchoides composticola* are the two important myceliophagous nematodes affecting mushroom crop. The extent of damage caused by these depends upon the nematodes speices present, initial density and the crop stage at which the nematodes gain access into the beds. The saprophagous nematodes cause less damage than myceliophagous type. The common nematodes belonging to this group include *Caenorhabditis elegans* and *Rhabditis* spp. Exclusion of these nematodes from gaining access through compost and casing is the best way of preventing nematode infestation. Chemical prophylactic control is also followed by use of nematicides.

CHAPTER 17

SOME IMPORTANT TIPS FOR GROWING MUSHROOMS SUCCESSFULLY

CONTENTS

P.K. Khanna (2003) Mushroom cultivations, Punjab Agricultural University, Ludhiana, Chapter X PP 55-56

Prevalence of unhygienic conditions in and around the mushroom farms and the meagre resources available with the mushroom growers are the major contributing factors towards crop failure. Sound sanitation measures and good hygienic practices are, therefore, very important for successfully growing mushroom crop. A few tips that should be routinely followed to get a reasonably good disease free crop *are* given below:

1. Growing rooms should have *pucca* floors and with a provision for cross-ventilation and good lighting arrangement, for watering, inspection and other connected operations.
2. Growing rooms should be white washed with lime and copper sulphate before use. As far as possible do not dry broom your mushroom house as well as its surroundings.
3. Composting should be done on *pucca* floor and proper dimensions of the stack (given in the text) should be followed.

4. Avoid leaching during wetting of straw/fertilizer mixture.
5. Lumps of compost, if any, formed during composting may be brocken.
6. Final grade compost should be checked for ammonia and adequacy of water in it. It should be free from ammonia and contain 68-72% moisture.
7. Excessive ammonia and water in the compost should be corrected by giving one or more turnings to the compost.
8. Healthy, contaminant free and fresh spawn should be used. Use 10 ml malathion in 10 litres of water for 100 sq.m. area as a precautionary measure against pests after spawning.
9. Casing materials should be processed on pucca floor preferably in-doors.
10. Casing soil (final grade material) should have good water holding and gaseous exchange capacity.
11. Wash hands with a disinfectant before casing.
12. Always apply moistened casing mixture to the beds.
13. As a precautionary measure, spray the cased beds with dithane Z-78 or benlate.
14. Unused casing (already disinfected/pasteurized) should be kept covered for its use in future.
15. Cover the holes, if any of the cased beds with fresh casing mixture.
16. Picking should be done by gently twisting the mushrooms with the hand. Do not harvest mushrooms with a knife.
17. The first choice (for picking) should be new mushroom houses, followed by old growing rooms.
18. Workers engaged in picking mushrooms should at least disinfect their hands and feet before entering growing houses as well as while going from one cropping room to another for harvesting mushrooms.
19. All the tools especially where disease exists should be thoroughly disinfected.
20. Clean the vacated containers with a stiff brush followed by thorough washing in water so that all the rubbish is removed.
21. Dispose off the spent compost as far away as possible from the mushroom house as possible.
22. Avoid the use of unknown chemicals in various operations of mushroom growing.
23. Maintain-complete record of the various operations of mushroom growing.
24. Do not leave anything to chance because prevention is always better than cure.

❑❑❑❑❑

CHAPTER 18

ECONOMICS OF MUSHROOMS

CONTENTS

P.K. Khanna and S. Kapoor (2007) Manual on mushroom production. Dept. of Microbilogy college of basic sciences and Humanities, PAU , Ludhiana PP 86-90

Introduction

The climate of Punjab is best suited for cultivation of 4 varieties of mushrooms namely the button mushroom, dhingri, milky mushroom and paddy straw mushroom. These mushrooms can be grown in different seasons of the year thus making it possible to have cultivation of mushrooms round the year. The state at present has more than 350 mushroom farms most of which are doing intensive cultivation of the button mushrooms during winter. However, there are few who grow dhingri and the straw mushroom as well, though at small scale. For convenience, these farms may be categorised as commercial, large scale and small scale farms.

Irrespective of the scale of cultivation, a grower must possess adequate knowledge of the growing technology and its economics. The economical aspects include the initial investments, cost of cultivation, sale proceeds and the net profits. A number of factors like labour cost, type of technology, cost of raw materials, energy input and fluctuations in the selling price of the produce in market determine the overall economics of the mushroom cultivation.

It is thought fit that mushroom cultivation at cottage scale is more appropriate for rural people as they have little or no land holdings. Consequently, small growing house of 1000 cu. ft. covered space has been considered as a unit to calculate the economics of mushroom growing. However, if scale of cultivation is increased, some of the costs get pooled and as such the cost of cultivation is considerably reduced.

Economics of white button mushroom at cottage scale.

Covered space: 10′ ×10′×10′ (length, breadth, height)

	Expenditure (Rs.)	
A. ONE CROP SYSTEM:		
Mushroom bed area 30 m^2		
Non-recurring:		
Wooden board 5′ × 5′, 3No.	1800	
Spray pump 1 No.	800	
Wooden boxes 150 No.	1200	
Forks, water pipe, buckets, balance, etc.	1000	
Miscellaneous	600	
Total	**5400**	**...(I)**
Recurring:		
Wheat straw, 8 qtls. @ Rs.150/qtl.	1200	
Fertilizers, additive, pesticides	700	
Spawn (@300g/sq.m.)	450	
Polythene bags 3kg (for packing mushrooms)	300	
Miscellaneous	200	
Total:	**2850**	**...(II)**
G. Total (I & II):	**8250**	
Depreciation of non-recurring items @ 30%	1620	...(III)
Returns (Rs.)		
Expected mushroom yield (225kg) Gross sale @ Rs.35/kg	7875	...(IV)
Net inputs (II & III)	4430	...(V)
Net profit (IV-V)	3400	
Cost of production/kg	19.86	
B. TWO CROP SYSTEM:		
Non-recurring expenditure for 2nd crop	Nil	
Recurring expenditure of two crops	5700	...(VI)
Net inputs (III & VI)	7320	...(VII)
Expected mushroom yieid(400kg) Gross sale	14000	...(VIII)
Net profit (VIII-VII)	6680	

Economics of dhingri at cottage scale

No. of crops: 3

Non-recurring	**Expenditure** (Rs.)	
Shelf system made of bamboo and polythene ropes	1200	
Spray pump	800	
Forks, water pipe, buckets, balance, etc.	1000	
Miscellaneous	600	
Total	**3300**	**...(I)**
Recurring:		
Polythene bags 24"x16", 240 No.	560	
Wheat straw, 5 qtls. @ Rs.150/qtls.	750	
Spawn 180 bottles	1980	
Packing bags	300	
Miscellaneous	210	
Total:	**3800**	**...(II)**
G.Total (I & II):		
Depreciation of non-recurring items @ 30%	990	...(III)
Returns (Rs.)		
Expected mushroom yield(275kg) Gross sale @ Rs.35/kg	9625	...(IV)
Net inputs (II & III)	4790	...(V)
Net profit (IV-V)	4835	
Cost of production/kg	17.41	

Economics of paddy straw mushroom at cottage scale.

Number of crops: 4

Non-recurring	**Expenditure (Rs.)**	
Bamboo sticks (3′, 16 pcs)	300	
Shelf system made of bamboo and polythene ropes	1200	
Total	**1500**	**...(I)**
Recurring:		
Paddy straw, 16 qtls. (only cartage)	500	
Gunny threads	150	
Spawn 40 bottles	440	
Bags for packing	70	
Total:	**1160**	**...(II)**

G.Total (I & II):		
Depreciation of non-recurring items @ 30%	450	...(III)
Returns (Rs.)		
Expected mushroom yield (140kg) Gross sale @ Rs.35/kg	4900	...(IV)
Net inputs (II & HI)	1610	...(V)
Net profit (IV–V)	3290	
Cost of production/kg .	11.50	

Economics of milky mushroom at cottage scale

No. of crops: 3

Non-recurring	**Expenditure (Rs.)**	
Shelf system made of bamboo and polythene ropes	1200	
Spray pump	800	
Forks, water pipe, buckets, balance, etc.	1000	
Miscellaneous	600	
Total	**3300**	**...(I)**
Recurring:		
Polythene bags **12" × 16"**, 700 No.	600	
Wheat straw, 5 qtls. @ Rs. 150/qtls.	750	
Spawn 200 bottles	2200	
Packing bags	300	
Fuel for boiling	200	
Miscellaneous	150	
Total:	**4200**	**...(II)**
G.Total (I & II):		
Depreciation of non-recurring items @ 30%	990	...(III)
Returns (Rs.)		
Expected mushroom yield(275kg) Gross sale @ Rs.35/kg	9625	...(IV)
Net inputs (II & III)	5190	...(V)
Net profit (IV-V)	4435	
Cost of production/kg	18.87	

❑❑❑❑❑

CHAPTER 19

MUSHROOM RECIPES

Mushrooms are best consumed as fresh. However, in actual practice this may not be possible. Their storage at high temperature results in brown discolouration. Mushrooms have a high rate of respiration and hence proper attention has to be paid in storage. In many countries where *Agaricus* marketing has been developed as a business they are covered by using special type of PVC film and transporting is done under refrigerated conditions at about 15°C. Under that condition it may keep well for about a week. They can be blanched also, (dipping in boiling water for 1 minute, drain and dry). Mushrooms can be packed in fibre board trays, covered with an inverted tray to reduce the desiccation, is known to have the maximum shelf life of 7 days at 1°C (Bano and Patwardhan, 1979). Vacum packing, use of nitrogen gas etc., are also some of the techniques to enhance shelf like. In recent times gamma radiation also offers a method of preservation.

Subtle in flavour and attractive in appearance, mushrooms lend themsleves to tasty culinary art. A number of tasty dishes can be prepared out of mushrooms with typical mushroom flavour. Fresher the mushrooms, shorter the cooking time.

Mushrooms should be prepared before cooking by cutting them into pieces of desired size, snipping the tips of stems and washing them thoroughly taking care to see that the water is drained completely.

1. Sauted Mushrooms

Mushrooms	:	150 gms.
Onion (medium sized)	:	One-cut to small pieces
Refined oil	:	Two table spoons

Salt and pepper powder to taste.

Heat the refined oil and add sliced onions. When half cooked add mushrooms, cover and cook over slow heat. Add a table spoon of water and if necessary another and cook till done. Sprinkle salt and pepper and serve hot with chapaties or sliced bread.

2. Creamed Mushrooms

Mushrooms	:	250 gms
Fat	:	Two and half table spoons
Maida	:	One table spoon
Milk	:	Three fourth cup

Salt and pepper powder to taste

Trim, wash mushrooms and remove excess water. Cook in fat until nearly tender. Add salt, pepper and maida mixing them well. Add milk gradually and simmer for 3 to 4 minutes and serve with hot buttered toast.

3. Cream of Mushroom Soup

Mushrooms	:	125 gms.
Butter	:	Two table spoons
Maida	:	Two table spoons
Milk	:	Two cups

Salt to taste

Trim, wash and chop mushrooms and cook them in half cup of water and a little milk. Set apart melted butter and blend in maida and salt. Add milk gradually constantly stirring until thickened. Add cooked mushrooms. Heat on a low flame and serve with chopped parsley.

4. Mushroom Sauce

Mushrooms	:	100 gms
Butter	:	Three table spoons
Maida	:	Two table spoons
Chicken	:	One cup

Salt and pepper powder to taste

Trim, wash and chop mushrooms. Cook mushrooms in one table spoon of butter. Keep them aside. Combine two table spoons of butter and maida and stir constantly and heat till cooked. Add cooked mushroom. Add salt and pepper powder and serve.

5. Mushroom Salad

Mushrooms	:	150 gms.
Shelled peas	:	200 gms.
Diced carrots	:	200 gms
Cubed potatoes	:	150 gms.
Refined oil	:	One table spoon

Pepper powder one pinch Salt to taste

Boil shelled peas, diced carrots and cubed potatoes in just enough water. Cook mushrooms in oil and add a table spoon of water. When cooked add the cooked peas, carrots and potatoes, toss and add salt and pepper. French dressing could also be used as an alternative to salt and pepper. Add two table spoons of salad oil or refined oil, two table spoons of vinegar, salt and pepper and half a spoon of sugar and toss them completely before serving.

6. Baked Mushrooms

Mushrooms	:	300 gms
Maida	:	Two and half table spoons
Milk	:	Three fourths cup
Refined oil'	:	Four table spoons

Salt to taste

Trim, wash and drain mushrooms thoroughly. Cook over a slow flame using two table spoons of refined oil and a table spoon of water. When three fourths cooked keep aside. Make a thick white sauce with maida and milk using the remaining oil. Add a few drops of the mushroom stock. Line a dish with the white sauce. Place cooked mushrooms over it. Place the rest of the sauce on the top of this. Bake for ten minutes in a pre-heated oven. Add salt to taste and serve.

7. Another recipe

Mushrooms	:	150 gms.
Onion	:	One large chopped
Butter	:	One fourth cup
Soft bread crumbs	:	One and half cup
Cashewnuts coaresly ground	:	Half a cup
Tomato ketchup	:	One table spoon
Lemon juice	:	One table spoon
Pepper powder	:	One table spoon

Salt to taste

Wash, trim and drain mushrooms thoroughly. Cook mushrooms in butter with chopped onions for about five minutes. Add soft bread crumbs, coarsely ground cashewnuts, tomato ketchup, lemon juice and pepper and cook for two more minutes. Transfer this to a baking dish and bake for twelve minutes in a moderate temperature. Add salt to taste and serve.

8. Mushroom ketchup (Sethi & Anand, 1978)

Fresh Mushrooms	:	500 gms	ground together
Cordomom	:	2 gms	
Pepper	:	2 gms	
Cloves	:	2 gms	
Mace	:	2 gms	
Cinnamon	:	2 gms	
Dry Ginger (Ground)	:	2 gms	
Red Chillies	:	2 gms	
Vinegar	:	500 ml	
Salt	:	12 gms	

Clean mushrooms and keep them in a clean porcelain or enamel bowl for 12 hours after sprinkling salt. Store the same in the vinegar for few days and blend the salted and soaked mushrooms to a slurry and add spices. Heat the mass till it thickens to the consistency of ketchup. Small quantity of meat extract or monosodium glutamate can also be added to enhance the flavour. The ketchup can be taken in bottles and they are sealed with corks and sterilised for 30 minutes; in boiling water. The bottles are cooled and kept in cool dry place.

9. Stuffed Capsicum Mushrooms

Mushrooms	:	350 gms.
Capsicums	:	large, four
Onion	:	Large one chopped
Fenugreek	:	One bunch
Green chillies	:	Two
Cooking oil	:	Three table spoons

Salt to taste

Wash and chop mushrooms. Wash and cut capsicums to stuff with mushrooms. Trim and remove seeds from capsicums cups. Replace caps and steam in a perforated vessel used for steaming for about ten minutes. Sprinkle salt and keep aside.

Heat oil, add chillies, onions and fenugreek leaves. When half done add chopped mushrooms. Cook under cover till mushrooms are done. Whem still hot, place mushrooms stuffing into capsicum cups. Replace caps and steam for two minutes. Remove, add salt to taste and serve.

10. Scooped Drumstick Mushrooms

Mushrooms	:	200 gms.
Fresh drumsticks (Boiled)	:	Eight
Green chillies	:	Two
Onion large	:	One
Roasted cumin powder	:	Half tea spoon
Bengalgram flour	:	One table spoon
Cooking oil	:	Two table spoons
Milk	:	Two table spoons

Salt to taste

Fry mushrooms quickly with two table spoons of oil and add one table spoon of milk. Remove pulp from the boiled drumsticks. Fry chopped onion in one table spoon of oil, and chillies, cumin powder and bengalgram flour. Add drumstick pulp and fried mushrooms. Add remaining milk and mix well. Add salt to taste and serve on toast.

11. Mushroom Omlet

Mushrooms	:	50 gms.
Egg yolks	:	Two
Cooking oil	:	Two table spoons

Salt and pepper powder to taste

Clean, cut and fry mushrooms quickly with one table spoon of oil and keep aside. Beat the egg whites until stiff. Beat the yolks until thick. Add two table spoons of hot water, salt and pepper to yolks. Mix the white and yellow yolks thoroughly. Add sliced mushrooms kept aside. Melt fat in an omlet pan, grease bottom and sides, turn over low head flame until it is puffy and light brown. Loosen edges of omlet, fold over twice and serve hot.

12. Mushroom Poached Eggs

Mushrooms	:	50 gms.
Fresh eggs	:	Two
Vinegar	:	One table spoon
Cooking oil	:	Two table spoons

Salt and pepper powder to taste

Clean, cut and fry mushrooms quickly in table spoon of oil. Keep aside. Heat half cup of water to simmering in a shallow pan. Add powdered salt and pepper and vinegar. Break eggs into a cup and slip carefully into water. Let the egg cook below boiling point in just simmering water for about five minutes, until white is firm and a film has formed over yolk. Add the fried mushrooms kept aside. Remove poached eggs and serve with toast.

13. Scrambled Egg Mushrooms with Coriander leaves

Mushrooms	:	150 gms.
Eggs	:	Four
Onion large	:	
Chopped coriander leaves	:	Two bundles
Big green chillies	:	Two
Finely chopped Curry leaves	:	Eight to ten
Medium sized tomato	:	One
Cooking oil	:	Two and half table spoons
Milk	:	Two table spoons

Clean and cut the mushroom, onion and chillies. Add oil to a pan and when hot add chillies, onions and curry leaves and mushrooms. Add one table spoon of milk and cook mushrooms under cover on a low flame. When done, add salt, tomatoes and coriander leaves and cook for about three more minutes. Break eggs over the pan and add the remaining milk stirring constantly. Remove when still moist and serve with bread or chapaties.

14. Mushroom Eggs

Mushrooms	:	250 gms.
Eggs	:	Two
Large onion	:	One
Fenugreek leaves (Methi bhaji)	:	Two bunches
Turmeric	:	One pinch
Cooking oil	:	Four table spoons
Milk	:	Three table -spoons

Curry powder and salt to taste

Heat cooking oil in a shallow pan. Add cut onions and a pinch of turmeric and a little curry powder. Add methi leaves and when half done add cleaned cut and washed mushrooms cut into pieces. Cover with lid and cook mushrooms in milk for 6 to 7 minutes till done. Beat eggs and gently pour over simmering mushrooms covering them completely. Cover with lid and cook till eggs get set. Lift edges with spatulla and carefully slide down pan eggs on mushrooms to a flat plate with egg side up. Serve hot.

15. Mushroom Flavoured Rice

Mushrooms	:	200 gms
Cleaned and washed rice	:	200 gms.
Clove	:	One
Cinnamon	:	One inch piece
Slit chillies	:	Two
Cooking oil	:	Six table spoons
One medium sized onion	:	chopped

Salt to taste

Heat oil in a pan. Add chillies, clove, cinnamon and onions. Add rice and fry for a minute. Add mushrooms and mix well. Add water and salt and cook till all the water is absorbed and rice cooked. This can be done in a pressure cooker. Serve hot.

16. Chicken and Mushroom Salad

Mushrooms	:	Half cup
Sour curd & Salad dressing	:	One cup
Diced cooked chicken	:	Two cups

Prepare sour curd salad dressing by mixing well one cup of sour curd with two table spoons of vinegar, one table spoon of lime juice, one tea spoon sugar, one tea spoon salt and chilli powder to taste. Beat the mixture until smooth. Add combined diced chicken and mushroom and toss. Serve cold. The cold chicken and mushroom salad is preferred by tourists on hot days.

17. Mushroom, Ham and Egg Sandwiches

Chopped creamed mushrooms	:	100 gms.
Ground and cooled ham	:	200 gms.
Hard boiled eggs chopped	:	Two
Butter bread slices	:	Ten to twelve
Butter	:	Two table spoons
Pepper powder	:	Half tea spoon

Salt to taste

Mix ham, eggs and creamed mushrooms. Add butter, salt and pepper powder and mix thoroughly. Spread on buttered slices of bread and serve.

18. Mushroom Pickles

Mushrooms	:	250 gm
Chilli powder	:	One tea spoon
Mustard powder	:	One tea spoon
Jeera powder	:	One tea spoon
Fenugreek seed powder	:	Half tea spoon
Turmeric powder	:	One tea spoon
Glacial acetic acid	:	Ten ml.
Refined oil	:	200 ml.
Salt	:	40 gms. (about 8 to 9 tea spoons)

Wash, trim and drain mushrooms thoroughly. Fry mushrooms quickly with 100 ml. of oil. Stir well and cook till three fourths of the water is evaporated and keep aside. Heat remaining oil and fry all spices stirring well for a minute. Add cooked mushrooms and let it simmer for ten minutes. Mix thoroughly. Store in clean dry jars.

19. Mushroom Pickles (other recipe)

Fresh mushrooms	:	125 gms.
Mustard powder	:	Half tea spoon
Turmeric powder	:	Half tea spoon
Chilli powder	:	Half tea spoon
Salt	:	Half tea spoon
Sesamum oil	:	Thirty ml.
Asfoteda	:	One pich

Boil mushrooms for ten minutes, drain water and dry in sunlight for two to three hours. Oil, spices and mushrooms are mixed well and filled in a clean dry jar. The jar is kept in sunlight for two to three hours and then stored in a dry place at room temperature.

20. Mushroom Pickle (Singh and Zakia Bano, 1977)

Mushrooms chopped	:	125 gms.
Cumin seed powder	:	Half tea spoon
Fenugreek seed powder	:	Half tea spoon
Coriander seed powder	:	Half tea spoon
Turmerc powder	:	Half tea spoon
Mustard seed pcwder	:	Half tea spoon
Green chillies	:	Five
Vinegar	:	45 ml.
Seasamum oil	:	45 ml.
Salt to taste		

Chop fresh mushrooms, wash in clean water twice and squeeze gently to remove excess water. Cook the prepared mushroom in 30 ml. sesamum oil with salt in a covered stainless steel pan on a simmering flame for 10 to 15 minutes and keep aside. Mix all the spices well. Slit green chillies and fry them slightly in about 15 ml. oil. Add spice mixture. When cool fill in clean dry jars and store in a dry place at room temperature.

Mushroom Products Standardised and Prepared by Action Rresearch Institute, Calcutta, (Bhowmik et el., 1992)

21. Mushroom Instant-soup

Contains mushroom powder, dehydrated milk, corn flour and other ingredients. Add 1 tea spoon of soup-powder to boiled water and stir to dissolve thoroughly for preparing two bowls of hot mushroom soup. Addition of steamed vegetables yields a pleasant taste and more nutrition. Prepare cold soup by pouring soup-powder and ice-cubes in ordinary water. Serve mushroom soup, hot or cold and with or without steamed vegetables, to all-time guests and make 'any-time is soup time'.

22. Mushroom Biscuit

Contains mushroom powder, wheat-four, soyabean powder and other ingredients. Use this protein biscuit as a crisp snack and serve it with or without tea, coffee, mushroom cordial or mushroom instant-soup.

23. Mushroom Chips (Vegetarian)

Contain mushroom, soyabean, maize and other cereals, pulses, grount-nut, salt and spices and are fried in refined groundnut oil. Unpack and bite and also help others to crunch the tasty snacks.

24. Mushroom Chips (Chicken-Blended)

Contain mushroom dehydrated chicken and other ingredients. Unpack and bite and or serve to your friends with pleasure.

25. Mushroom-Garlic Sauce

Contains mushroom, garlic, onion and other ingredients. Use as a delicacy and an effective taste maker.

26. Mushroom Flakes (Vegetarian)

Contain mushroom, soyabean, maize and other cereals, pulses, ground-nut, salt and spices. Fry and serve as savoury snacks with hot or cold drinks which to suit your taste, may be soft or hard. Use as a good appetizer and a delicious gritty food.

27. Mushroom Flackes (Chicken-Blended)

Contain mushroom, chicken and other ingredients. Fry and serve with or without drinks.

28. Mushroom Cooking Powder

Contains mushroom powder and powder of all important Indian spices. Add this to increase taste and viscosity while preparing curry of mushroom, fish, meat, egg or vegetable.

Note. Mushroom growers should obtain a Licence under the Fruit Products Order (FPO), Govt. of India for the manufacture of Mushroom Products for sale. Licences under the order issued by

Deputy Director (F & VP), Ministry of Food Processing
Govt. of India, Shastri Bhavan
Madras-600 006.

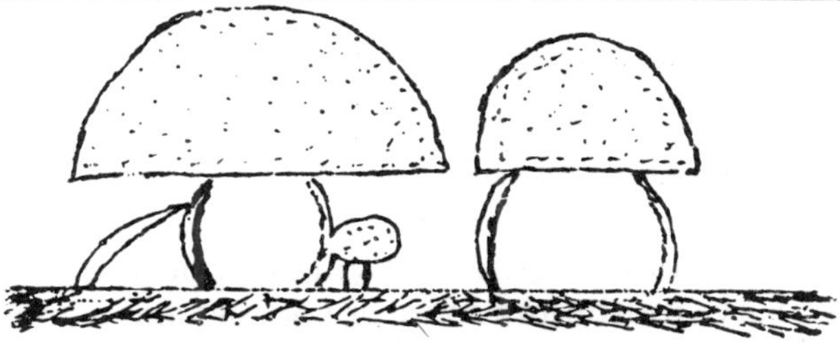

Grow Mushrooms for More Money

BIBLIOGRAPHY

Asthana, R.P. (1947). Mushroom Cultivation in Central Provinces. Magazine Agriculture College, Nagpur 22:26-31.

Bano, Z. & Patwardhan, M.V. (1979). Post harvest technology of Mushrooms. National Seminar on Mushrooms ICAR.

Bhowmik, K.L., Sinha, S., Shekhara, M.T.C. (1992). Mushroom Spawn Production Manual, Action Research Institute, Calcutta-700 004.

Bose, S.R. (1921). Possibilities of mushroom industry in India by cultivation, Agriculture Journal of India 16, 643-7.

Central Food Technological Research Institute, (1990). Mysore, India Home Scale Processing and Preservation of Fruits and Vegetables.

Central Food Technological Research Institute, (1989). Mysore, Profiles on Selected Technologies for Food and Agro Based Industries.

Dickinson Colin and John Lucas (1979). The encyclopedia of mushroom, (edited) orbis publishing, London, PP 280.

Gupta, G.K., Bajaj, B.S. and Suryanarayana, D. (1970). Studies on the cultivation of paddy-straw mushroom in India. Indian Phytopathology 23:621-51.

Indian Institute of Horticultural Research (1986). Mushroom Cultivation in Extension Bulletin No. 8,5-36, Bangalore.

Jandaik, C.L. and Kapoor, J.N. (1976). Studies on the cultivation of *P. sajorcaju* (*Ft.*) Singer. Mushroom Science 9(1): 667-72.

Jandaik, C.L. (1976). Commercial Cultivation of P. sajorcaju, Indian Journal of Mushrooms 2,19-24.

Kapoor, T.N. & Nita Behl (1982). Mushroom Growing for Beginners Division of Mycology & Plant Pathology, IARI, Ney Delhi pp. 72.

Munjal, R.L. (1973). Production of quality spawn of Agaricus bisporus and Volvariella species. Indian Journal of Mushrooms 1(1): 1-4.

Padwick, G.W. (1941). Mushroom cultivation in India. Indian Farming 11:363-6.

Prakash, T.N.Tejaswini and Ramana, R. (1986). Mushroom a Promising Crop for Future, SBM Farm News, 9, 5-7.

Prakash, T.N.and Tejaswini (1991). Prospects of Mushroom Cultivation in India. Proceedings of the National Symposium on Mushroom organised by the Kerala Agricultural University in collaboration with the Mycological Society of India, pp. No. 259-262.

Purkayastha, R.P., Biswas, S. and Rao, A.K. (1986). Some observations on the cultivation of Volvariella volvacea in the plains of West Bengal. Indian Journal of Mushroom 6:1-9.

Rajarathnam, S.Bano, Z. & Singh, N.S. (1983). Efficacy of formaldehyde and liquor Ammonia for controlling the growth of Sclerotium rolfsil during cultivation of the mushroom Pleurotus fiabellatus. Mushroom News Letter for the tropics 3. 3-10.

Ramakrishnan, &., Lalitha Kumari, D., Shanmugam, N. and Krishnamurthy, C.S., (1968). A simple technique for increasing the yield of straw mushroom, Volvariella displasia (Berk. & Br.) Sacd. Madras Agricultural Journal 55; 194-5.

Rath, C.C. (1961). A trial on the cultivation of paddy-straw mushroom at Banthra Nursery. Science and Culture 27: 207-8.

Seth, P.K. (1976). Supplementation of organic substrates of synthetic composts for increased mushroom yields, Indian Journal of Mushroom 2:36-39.

Sethi, V. and Anand, J.C. (1978). Processing of Mushrooms, Indian Mushroom Science : 1, 232-238.

Singh, N. S., and Zakia Banu (1977). Indian Food Packer Volume XXXI No: 5 pp. No. 18-19.

Su, U.T. and Seth, L.N. (1940). Cultivation of straw mushroom. Indian Farming 1:332-3.

Swaminathan, M.W. (1982). Indian Agriculture at the Cross-roads in Current Science 51:13-24.

Thomas, K.M. Ramakrishnan, T.S. and Narasimhan, I.L. (1943). Paddy-straw mushroom. Indian Farming 1:332-3.

Treshchow, C. (1944). Nutrition of the Cultivation Mushroom. Dansk botanisk Arkiv 11:1-180.

Vijay, B. Sohi, H.S. & Poadhyay, R.C. (1985). Control of green mould Trichoderma viride by use of Bavistin in the cultivation of Pleurotus sajor-caju (Fr.) Sing. Indian Phytopathology 39, 159.

Vijay, B. & Sohi, H.S. (1987). Cultivationof Oyster Mushroom Pleurotus sajor-caju (Fr.) Singer on chemically sterilized wheat straw. Mushroom Journal of Tropics, 7, 67-75.

Vijay, B. & Sohi, H.S. (1989). Fungal competitors of Pleurotus sajor-caju (Fr.) Sing. Mushroom Journal of Tropics, 9,29-35.

Vijhayakhader and B. Nayana Pandya (1981 a). Nutritional Studies on Paddy Straw Mushroom. Indian Journal of Mushrooms. VII (1 &2)p. 18-25.

Vijhayakhader and B. Nayana Pandya (1981 b). Acceptability Studies on Weaning Foods and Pickle Prepared out of paddy straw Mushrooms and the keeping quality of the same. Ind. J. Mushrooms VII (1 & 2) p. 31-36.

Vijhayakhader (1988). Studies on Oyster Mushrooms (Pleurotus). The Andhra Agril. Journal Vol. 35 (3 & 4) p. 212-214.

Zeayen, D.V., (1972). Agricultural Research Report, 782, p.130.

PUBLICATIONS ON MUSHROOMS BY VIJAYAKHADER

1. Button mushrooms for high proteinous diet
2. Studies on Oyster Mushroom
3. Nutritional Studies on Paddy straw Musroom
4. Acceptability Studies on weaning Foods and Pickles
5. Impact of training on cultivation of mushrooms to the unemploed
6. Suitable technology for cultivation of mushroom in small farms
7. Rural Women as Entrepreneurs in Mushroom Cultivation
8. Rural Women-nutritional security
9. Mushrooms for Many uses Food & Nutrition
10. Effect of fiber from Mushrooms, Spinach and Isabgol on blood glucose response in Non-Insulin Dependent Diabetics
11. Hypolipidemic action of Volvariella in albino rats

OTHER PUBLICATIONS

12. Mushroom cultivation offers lucrative business
13. Edible oyster mushrooms on cotton plant stalks

BUTTON MUSHROOM FOR HIGH PROTEINOUS DIET

Vijaya Khader (1996) Button mushroom to high proteinous diet Indian Farming, Oct. 1996, Special issue on world Food Day, 45–51

Vijaya Khader

It takes approximately 15 weeks to complete an entrie production cycle, from the start of composting to the final steaming off after harvesting has ended. Final yield depends on how well a grower has monitored and controlled the temperature, humidity, pests, and so on. All things considered, the most important factors for good production appear to be experience plus an intutive feel for the biological rhythms of the commercial mushroom. The production system used to grow a crop can be chosen after the basics of mushroom growing are understood.

Cultivation of white button mushroom in South India started during 1971-72. During the recent years the commercial cultivation of mushrooms in India is gaining momentum as one of the significant export-oriented agro-industrial enterprises. The Government of India has been emphasizing the importance of agro-based exports due to the advantages of rural employment generation and gainful utilization of material and farm resources, At present there is an appreciable demand for the white button mushroom. This mushroom requires a higher initial investment and quality control compared with other varieties. It would be advisable to produce this variety for earning foreign exchange.

Prof. (Mrs.) Vijaya Khader participated in a Work Shop on Mushrooms at Kerala Agricultural University.

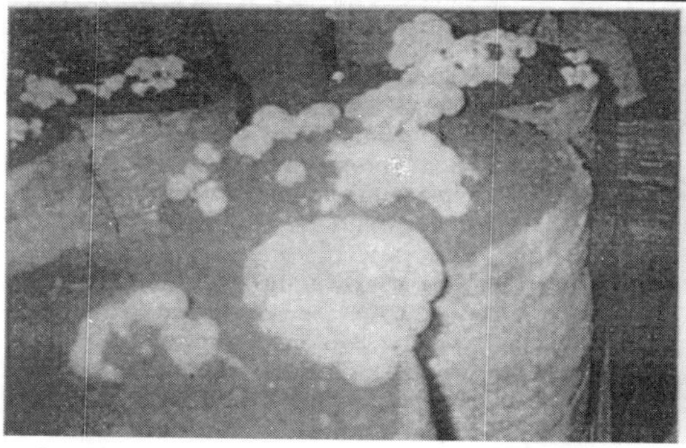

Gorowing button mushroom on small scale

Growing button mushroom on large scale

Botton mushroom (*Agaricus bisporus*) farming consists of 6 steps, Although the divisions come arbitrarily, these steps identify what is needed to form a mushroom production system. The 6 steps are:

1. Phase I composting
2. Phase II composting
3. Spawning
4. Casing
5. Pinning
6. Cropping

These steps are described in their naturally occurring sequence, emphasizing the salient features within each step. Two types of materials are generally used for mushroom compost, the most used and least expensive being wheat straw-bedded horse manure. Synthetic compost is usually made from hay and crushed corn, cobs, although the term often refers to any mushroom compost where the prime ingredients is not horse manure. Both types of compost require the addition of nitrogen supplements and a conditioning agent, gypsum.

The preparation of compost occurs in 2 steps, Phase I and Phase II composting. The discussion of compost preparation and mushroom production begins with Phase I composting.

PHASE I : MAKING MUSHROOM COMPOST

This Phase usually occurs outdoors although an enclosed building or a structure with a roof over it may be used. A concrete slab is required for composting. In addition, a compost turner to aerate and water the ingredients and a tractor-loader to move the ingredients to the turner is needed. In earlier days piles were turned by hand using pitchforks, which is still an alternative to mechanized equipment, but is labour intensive and physically demanding. Phase I composting is initiated by mixing and wetting the ingredients as they are stacked in a rectangular pile with tight sides and a loose centre. Normally, the bulk ingredients are put through a compost turner. Water is sprayed onto the horse manure of synthetic compost as these materials move through the turner. Nitrogen supplements and gypsum are spread over the top of the bulk ingredients and are thoroughly mixed by the turner. Once the pile is wetted and formed, aerobic fermentation (composting) commences as a result of the growth and reproduction of micro-organisms which occur naturally in the bulk ingredients. Heat, ammonia and carbon dioxide are released as by-products during this process. Compost activators, other than these mentioned, are not needed, although some organic farming books stress the need for an 'activator'.

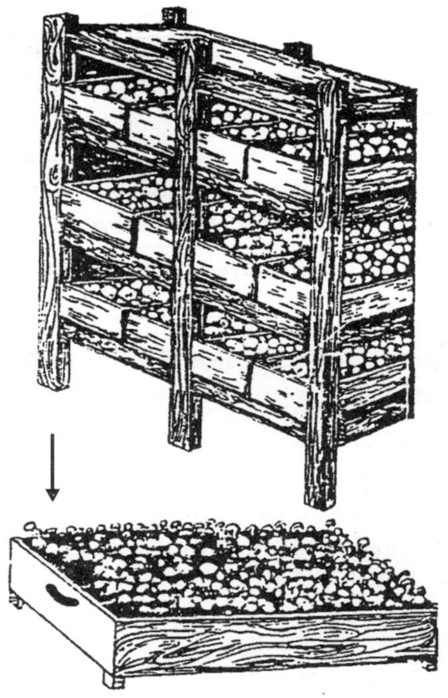

Mushroom growing rack

Mushroom compost develops as the chemical nature of the raw ingredients is converted by the activity of micro-organisms heat, and some heat-releasing chemical reactions. These events result in a food source most suited for the growth of the mushroom to the exclusion of other fungi and bacteria. There must be adequate moisture, oxygen, nitrogen and carbohydrates present throughout the process or else the process will stop. This is why water and supplements are added periodically, and the compost pile is aerated as it moves through the turner.

Gypsum is added to minimize the greasiness compost normally tends to have. Gypsum increases the flocculation of 'certain chemicals in the compost, and they adhere to straw or hay rather than filling the pores (holes) between the straws. A side benefit of this phenomenon is that air can permeate the pile more readily; and air is essential to the composting process. The exclusion of air results in an airless (anaerobic)

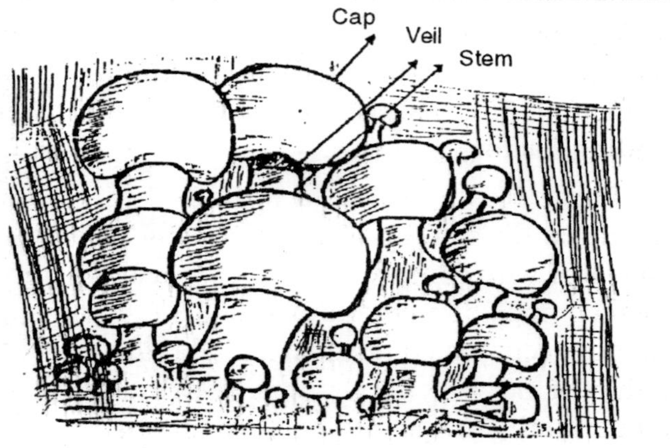

Mature mushrooms of different sizes

environment in which deleterious chemical compounds are formed which detract from the selectivity of mushroom compost for growing mushrooms. Gypsum is added at the outset of composting at 18.14 kg/ton of dry ingredients.

Nitrogen supplements in general use today include brewer's grain,seed meals of soybeans, peanuts or cotton and chicken manure, among others. The purpose of these supplements is to increase nitrogen content to 1.5% for horse manure or 1.7% for synthetic, both computed on a dry weight basis. Synthetic compost requires the addition of ammonium nitrate or urea at the outset of composting to provide the compost microflora with a readily available form of nitrogen for their growth and reproduction.

Corn cobs are sometimes unavailable or available at a price considered to be excessive. Substitutes for or complements to corn cobs include shredded hardwood bark, cotton seed hulls, neutralized grape pomace, and cocoa bean hulls. Management of a compost pile containing any one of these materials is unique in the requirements for watering and the interval between turnings.

The initial compost pile should be 1.52 m–1.82 m wide, 1.52 m–1.82 m high and as long as is necessary. A 2-sided box can be used to form the pile. The sides of the pile should be firm and dense, yet the centre must remain loose throughout Phase I composting. As the straw or hay softens during composting, the materials become less rigid and compaction can easily occur. If the materials become too compact, air cannot move through the pile and an anaerobic environment will develop.

Turning and watering are done at approximately 2-day intervals, but not unless the pile is hot (62.8°C–76.7°C). Turning provides the opportunity to water, cerate and mix the ingredients, as well as to relocate the straw of hay from a cooler to a warmer area in the pile, outside vs inside. Supplements are also added when the piles are turned, but they should be added early in the composting process. The number of turnings and the time between turnings depend on the condition of the starting material and the time necessary for the compost to heat to temperatures above 62.8°C.

Water addition is critical since too much will exclude oxygen by occupying the pore space, and too little can limit the growth of bacteria and fungi. As a general rule, water is added up to the point of leaching when the pile is formed and at the time of first turning and thereafter either none or only a little is added for the duration of composting. On the last turning before Phase II composting, water can be applied generously so that when the compost is tightly spueezed, water drops from it. There is a link between water, nutritive value, microbial activity and temperature, and because it is a chain, when one condition is limiting for one factor, the whole chain will cease to function. Biologists see this phenomenon repeatedly and have termed it the Law of Limiting Factors.

Phase I composting lasts from 7 to 14 days, depending on the nature of the material at the start and its characteristics each turn. There is a strong ammonia odour associated with composting, which is usually complemented by a sweet, mouldy smell. When compost temperatures are 68.3°C and higher and ammonia is present, chemical changes occur which result in a food rather exclusively used by the mushroom. As a by-product of the chemical changes, heat is released and the compost temperatures increase. Temperatures in the compost can reach 76.7°C–82.2°C during the second and third turnings when a desirable level of biological and chemical activity is occurring. At the end of Phase I the compost should (a) have a chocolate brown colour; (b) have soft, pliable straws; (c) have a moisture content of 68-74%; and (d) have soft, pliable smell of ammonia. When the moisture, temperature, colour and odour described have been reached, Phase I composting is completed.

PHASE II: FINISHING THE COMPOST

There are 2 major purposes to Phase II composting. Pasteurization is necessary to kill insects, nematodes, pest, fungi, or other pests that may be present in the compost. And second, it is necessary to remove the ammonia which formed during Phase I composting. Ammonia at the end of Phase II in a concentration higher than 0.07% is often lethal to mushroom spawn growth, thus it must be removed, generally a person can smell ammonia when the concentrations above 0.10%. Phase II takes place in one of the 3 places, depending on the type of production system used. For the zoned system of growing, compost is packed into wooden trays, the trays are stacked 6-8 high and are moved into an environmentally controlled Phase II room. Thereafter, the trays are moved to special rooms, each designed to provide the optimum environment for each step of the mushroom growing process. With a bed or shelf system the compost is placed directly in the beds.; which are in a room used for all steps of crop culture. The most recently intoduced bulk system, is one in which the compost is placed in a cement block bin with a perforated floor and no cover on top of the compost; this room is specifically designed for Phase II composting.

The compost, whether placed in beds, trays or bulk, should be filled uniformly in depth and density or compression. Compost density-should allow for gas exchange, since ammonia and carbon dioxide will be replaced by outside air.

Phase II composting can be viewed as, a controlled temperature dependent, ecological process using air to maintain the compost in a temperature range best suited for the deammonifying organism to grow and reproduce. The growth of these thermophilic (heat–loving) organisms depends on the avilability of usable carbohydrates and nitrogen, some of the nitrogen in the form of ammonia.

Optimum management for Phase II is difficult to define and most commercial growers tend toward one of the 2 systems in general use today, high temperature or low temperature. A high temperature Phase II system involves an initial pasteurization period during which the compost and the air temperatures are raised to atleast 62.8°C for 6 hr. This can be accomplished by heat generated during the growth of naturally occurring micro-organisms, or by injecting steam into the room where the compost has been placed, or both. After pasteurization the compost is reconditioned by immediately lowering the temperature to 60.0°C by flushing the room with fresh air. Thereafter, the compost is allowed to cool gradually at a rate of approximately -16.7°C to -16.1°FC each day until all the ammonia is dissipated. This Phase II system requires approximately 10-14 days to complete.

In the low temperature Phase II system the compost temperature is initially increased to about 52.2°C with steam or by the heat released via microbial growth, after which the air temperature is lowered so the compost is in a temperature range of 51.7°C–54.4°C. Pasteurization is intiated 24-48 hr later by injecting steam into the room, which raises the air and compost temperatures to 60.0°C. These temperatures are monitored and maintained from 2–4 hr, which completes the pasteurization requirement for the compost. After pasteurization, the air temperature is lowered to force the compost temperature back to the 51,7°C–54.4°C range. During 4–5 days after pasteurization, the compost temperature may be lowered by about -16.7°C a day until the ammonia is dissipated. In is important to remember the purposes of Phase II when trying to determine the proper procedure and sequence to follow, one purpose is to remove unwanted ammonia. To this end the temperature range from 51.7°C–54.4°C is most efficient since deam-monifying organisms grow well in this temperature range. A second purpose of phase II is to remove any pest present in the compost by use of a pasteurization sequence.

At the end of Phase II the compost temperature must be lowered to approximately 23.9°C-26.7°C before spawning (planting) can begin. The nitrogen content of the compost should be 2.0-2.4%, and the moisture content between 68 and 72%. Also, at the end of the Phase II it is desirable to have 2.26 kg-3.17 kg of dry compost/31 cm^2 of bed or tray surface to obtain profitable mushroom yield. It is important to have both the compost and the compost temperatures uniform during the Phase II process since it is desirable to have as homogenous a material as possible.

Increasing food production to meet our country's needs is essential. with the existing conventional agricultural products and methods, we may not be able to cope up with the food problem. There is an urgent need to identify, cultuvate and popularize new foods which cannot only augment the food deficits but also are nutritious and easily available.

Mushrooms are good supplements for protein lacking diets and can be easily cultivated indoors and marketed profitably. They are a fair source of vitamin C, niacin, pantothenic acid, folic acid, calcium, phosphorus and iron. Due to negligible amount of carbohydrate and fat mushroom serve as a good diet for diabeties and those who suffer from excess fat in their bodies.

During the recent years, the commerical cultivation of mushrooms in India is gaining momentum as one of the significant export-oriented agro-industrial enterprises due to the advantage of rural employment genration and gainful utilization of natural and farm resources. Mushroom cultivation is emerging in India, as a promising agro-based, land-independent enterprise due to filexibility in its operpation. The commerical cultivation of mushrooms in India is a recent origin and is restricted to 3 varieties, namely white button mushrooms (*Agricus bisporus*), paddy straw mushroom (*Valverella sp.*) and oyster mushroom (*Pluerotus sp.*). Mushroom cultivation is labour-intensive. About 2 250 mandays of employment are generated in production of 9, 450 kg of white button mushroom. About 1,000 mandays are required to produce 4,800 kg of oyster mushroom, whereas 50 mandays to produce 360 kg of paddy straw mushrooms.

Market trends indicate the cosumer preferences towards white button mushrooms which relaize higher price than other varieties. It may be advisable to produce this variety for earning foregin exchange. Button mushroom cultivation requires higher investment and optimum temperature of 16°C.

SPAWNING

Mushroom compost must be inoculated with mushroom spawn, if one expects mushrooms to grow. The mushroom itself is the fruit of a plant as tomatoes are of tomato plants. Within the tomato one finds seeds, and these are used to start the next season's crop. Microscopic spores form within a mushroom cap, but their small size precludes handling them like seeds. As the tomato comes from a plant with roots, stems, and leaves, the mushroom arises from thin, thread-like cells called mycelium. Fungus mycelium is the white, thread-like plant often seen on rotting wood or mouldy bread. Mycelium can be propagated vegetatively, like separating daffodil bulbs and getting more daffodil plants. Specialized facilities are required to propagate mycelium, so the mushroom mycelium does not get mixed with the mycelium of other fungi. Mycelium propagated vegetatively is known as spawn, and commercial mushroom farmers purchase spawn from spawn companies.

Spawn makers start the spawn making process by sterilizing a mixture of *rye* grain plus water and chalk; wheat, millet and other small grain may be substituted for *rye.* Sterilized horse manure formed into blocks was used as the growth medium for spawn up to about 1940 and this was called block or brick spawn or manure spawn, such spawn is uncommon now. Once sterilized grain has a bit of mycelium added to it, the grain is incubated at 23.3°C and the bottle containing the grain and mycelium is shaken

3 times at 4-day intervals over a 14-day period of active mycelial growth. Once the grain is colonized by the mycelium, the product is called spawn. Spawn can be refrigerated for a few months, so spawn is made in advance for farmer's order.

Spawn is distributed on the compost and then thoroughly mixed into the compost. For years this was done by hand, broadcasting the spawn over the surface of the compost and ruffing it with a small rake-like tool. In recent years, however, for the bed system, spawn is mixed into the compost by a special spawning machine which mixes the compost and spawn with tines or small finger like devices. In a tray or batch system, spawn is mixed into the compost as it moves along a conveyor belt or while falling from a conveyor into a tray. The spawning rate is expressed as a unit; I unit per 3.04 m is desirable. The rate is sometimes expressed on the basis of weight vs compost weight; a 2% spawning rate is desirable.

Once the spawn has been mixed throughout the compost and the compost worked so the surface is level, the compost temperature is maintained at 23.9°C and the relative humidity is kept high to minimize drying of the compost surface or the spawn. Under these conditions the spawn will grow-producing a thread like network of mycelium throughout the compost. The mycelium grows in all directions from a spawn grain, and eventually the mycelium from the different spawn grains fuse together, making a spawned bed of compost one biological entity. The spawn appears as a white to blue-white mass throughout the compost after fusion has occurred. As the spawn grows it generates heat, and if the compost temperature increases to above 26.7°C–29.4°C, depending on the cultivar, the heat may kill or damage the mycelium and eliminate the possibility of maximum crop productivity and/or mushroom quality. At temperatures below 23.3°C spawn growth is slowed and the time interval between spawning and harvesting is extended.

The time needed for spawn to colonize the compost depends on the spawning rate and its distribution, the compost moisture and temperature, and the nature or quality of the compost. A complete spawn run usually requires 14–21 days. Once the compost is fully grown with spawn, the next step in production is at hand.

CASING

Casing is a top-dressing applied to the spawn-run compost on which the mushrooms eventually form. Clay-loam field soil, a mixture of peat moss with ground limestone, or reclaimed weathered, spent compost can be used as casing. Casing does not need nutrients since it acts as a water reservoir and a place where rhizomorphs form. Rhizomorphs look like thick strings and form when the very fine mycelium fuses together. Mushroom initials (pins), pri-mordia or pins form on the rhizomorphs, so without rhizomorphs there will be no mushrooms. Casing should be pasteurized to eliminate insects and pathogens it may be carrying. Also, it is important that the casing be distributed so the depth is uniform over the surface of the compost. Such uniformity allows the spawn to move into and through the casing at the same rate and ultimately for mushrooms to develop at the same time. Casing should be able to hold moisture since miosture is essential for the development of a firm mushroom.

Managing the crop after casing requires that the compost temperature be kept at around 23.9°C for up to 5 days after casing and the relative humidity should be high. Thereafter, the compost temperature should be lowered about –16.7°C each day until small mushroom initials have formed. Throughout the period following casing, water must be applied intermittently to raise the moisture level to field capacity before the mushroom pins form. Knowing when, how and how much water to apply to casing is an art which readily separates experienced growers from beginners.

PINNING

Mushroom initials (pins) develop after rhizomorphs have formed in the casing. The initials are extremely small but can be seen as outgrowths on a rhizomorph. Once an initial quadruples in size, the structure is a pin. Pins continue to expand and grow larger through the button stage and ultimately a button enlarges to a mushroom. Harvestable mushrooms appear 18–21 days after casing. Pins develop when the carbon dioxide content of room air is lowered to 0.08% or lower, depending on the cultivar, by introducing fresh' air into the growing room. Outside air had a carbon dioxide content of about 0.04%.

The timing of fresh air introduction is very important and is learned only through experience. Generally, it is best to ventilate as little as possible until the mycelium has begun to show at the surface of casing. It is also important to keep the relative humidity high by watering the walls and floor, while the mycelium grows through the casing and to stop watering at the time when pin initials are forming. If the carbon dioxide is lowered too early by airing too soon, the mycelium stops growing through the casing and mushroom initials form below the surface of the casing. As such mushrooms continue to grow, they push through the casing and are dirty at harvest time. Too little moisture can also result in mushrooms forming below the surface of the casing. Pinning affects both the potential yield and quality of a crop and is a significant step in the production cycles.

CROPPING

The terms flush, break, or bloom are names given to the repeating 3-5 days harvest periods during the cropping cycle; these are followed by a few days when no mushrooms are available to harvest. This cycle repeats itself in a rhythmic fashion, and harvesting can go on as long as mushrooms continue to mature. Most mushroom farmers harvest for 35–42 days, although some harvest a crop for 60 days, and harvest can go on for as long as 150 days.

The air temperature during cropping should be held between 13.9°C and 16.7°C for good results. This temperature range not only favours mushroom growth, but cooler temperatures can lengthen the life cycles of both disease pathogens and insect pests.

It may seem odd that there are pests which can damage mushrooms, but no crop is grown that does not have to compete with other organisms. Mushroom pests can cause total crop failures, and often the deciding factor on how long to harvest a crop is based on the level of pest infestation. These pathogens and insects can be controlled by cultural practices coupled with the use of pesticides, but it is most desirable to exclude these organisms from the growth rooms.

The relative humidity in the growing rooms should be high enough to minimize the drying of casing but not so high as to cause the cap surfaces of developing mushrooms to be clammy or sticky. Water is applied to the casing so water stress does not hinder the developing mushrooms; in commercial practice this means watering 2–3 times each week. Each watering may consist more of fewer gallons, depending on the dryness of the casing, the cultivator being grown, and the stage of development of the pins, buttons or mushrooms. Most first time growers apply too much water and the surface on the casing. Sealed casing prevents the exchanged of gases essential for mushroom pin formation. One can estimate how much water to add after first break has been harvested by realizing that 90% of the mushroom is water and a gallon of water weighs 4.54 litres(Imp). If 45.35 kg of mushrooms were harvested, 49.94 litres (Imp) of water were removed from the casing, and this is what must be replaced before second break mushrooms develop.

Outside air is used to control both the air and compost temperatures during the harvest period. Outside air also displaces the carbon dioxide given off by the growing mycelium. The more mycelial growth, the more carbon dioxide produced, and since more growth occurs early in the crop, more fresh air is needed during the first 2 breaks. The amount of fresh air also depends on the growing mushrooms, the area of the producing surface, the amount of compost in the growing room, and the condition or composition of the fresh air being introduced. Experience seems to be the best guide regarding he volume of air required.

A question frequently arises concerning the need for illumination while the mushrooms grow. Mushrooms do not require light to grow, only green plants require light for photosynthesis. Growing rooms can be illuminated to facilitate harvesting or cropping practices, but it is more common for workers or mushroom farmers to be furnished with miner's lamps rather than illuminating an entire room.

Ventilation is essential for mushroom growing, and it is also necessary to control humidity and temperature. Moisture can be added to the air by a cold mist or by live steam, or simply by witting the walls and floors. Moisture can be removed from the growing room by (i) admitting a greater volume of outside air, (ii) introducing drier air; and (iii) moving the same amount of outside air and heating it to a higher temperature since warmer air holds more moisture and thus lowers the relative humidity. Temperature control in a mushroom growing room is no different from temperature control in your home. Heat can originate from hot water circulated through pipes mounted on the walls. Hot, forced air can be blown through a ventilation duct, which is rather common at more recently built mushroom farms.

Mushrooms are harvested in a 7–10 day cycle, but this may be longer or shorter depending on the temperature, humidity, cultivar, and the stage when they are picked. When mature mushrooms are picked, an inhibitor to mushroom development is removed and the next flush moves towards maturity. Mushrooms are normally picked at a time when the veil is not too far extended. Consumers in North America want closed, tight, mushrooms while in England and Australia open, flat mushrooms are desired. The maturity of mushroom is assessed by how far the veil is streched, and not by how large the mushroom is. Consequently, mature mushrooms are both large and small, although farmers and consumers alike prefer medium to large size mushrooms.

Picking and packaging methods often vary from farm-to-farm. Freshly harvested mushrooms must be kept refrigerated at 1.7°C–7.2°C. To prolong the shelf-life of mushrooms, it is important that mushrooms 'breathe' after harvest, so storage in a non-waxed paper bag is preferred to a plastic bag.

After the last flush of mushrooms has been picked, the growing room should be closed off and the room pasteurized with steam. This final pasteurization is designed to destroy pests which may be present in the crop or the woodwork in the growing room, thus minimizing the likelihood of infesting the next crop.

There are more than one and a half million species of fungi which have enormous potential-valuable as food, antibodies, producers of enzymes, etc, and can be used in industries as pollution and insect controllers. On the other hand, they are responsible for the rejection of organ implants in human beings, cause secondary ailment in AIDS, thrive in oils and attack the fuel injection systems of jet planes attack and destroy paintings in no time. The Victorians collected mushrooms in the forest for food.

During the last several years, fungi have been extensively used as food. RHM (Rank Hovis McDougall), the largest food company in England, introduced myco-proteins based on fungi. Myco-proteins are extensively used in biscuits and ICI's animal feed, Pruteen, contains a fungal protein similar to mushrooms. RHM ferments a fungi *Fusarium graminearum* and harvests the biomass for further treatments. This is turned into a food-like material which tastes similar to soy protein. RHM perfected ways of enhancing the texture of myco-protein. You can make vegetarian sausage, steak or burger, add other flavour enhancers. It has a protein content of 47% against 57% protein of beef. The fibre content is almost like whole meal bread and there is no problem with cholesterol. The mushrooms, which we use, are another fungi widely used. Wild mushrooms are big business in the US since the National Park Law has outlawed the picking of wild mushrooms from the forest. The more we study fungi the more uses they can be put to.

□□□□□

STUDIES ON OYSTER MUSHROOM (PLEUROTUS)

The Andhra agric. J.(1998) 35 (3 & 4) 212-214,

Vijaya Khader

ABSTRACT

Nutrtitive value of gray and white varieties of Oyster Mushroom, cultivated under rural conditions was studied. An average yield of 2 kg bed–1 was obtained in three pickings for gray variety. Addition of a little of legume seed flour along with the spawn was found to increase the size and yield of the mushrooms.

Mushrooms are excellent sources of vitamins and minerals. They contain appreciable amounts of miscin, pantothonic acid and biotin (Subramanyan, 1986). In general, fresh mushrooms were found to contain 85-90% moisture, 2-4% protein, 4% carbohydrate, 0.3-0.4% fat and 1% minerals and vitamins (Shoba Davi and Sarojini, 1983). Although few studies were carried out on the paddy straw mushroom (*Volvariella valvaccea*) (Vijayakhader and Narayana, 1981), no systematic study was made on the cultivation and nutritive value of Oyster mushroom, which grows extensively in tropics. An attempt was therefore made in the present investigation to cultivate oyster mushrooms in rural areas and predict their nutritional quality.

MATERIAL AND METHODS

Oyster mushrooms of white (*Pleurotus citrino pileatus*) and gray (*Pleurotus sajor caju*) varieties were cultivated on paddy straw beds, during 1987 at two villages viz., Hanumanthanagar and Nandirajuthota near Bapatla (AP) using the spawn obtained from the department of Plant Pathology, Tamilnadu Agricultural University, Coimbatore. The beds were prepared by chopping one kg of paddy straw into 3-4 cm bits, soaked in cold water for 12-16 hr and the excess water was drained off. The straw bits were then dipped in hot water (80-85°C) for 20-30 mts., removed, cooled to room temperature under shade and filled in polythene bags (60 × 30 cm) size and 70-100 gauge thick) in alternate layers with little spawn and a pinch of legume seed powder (green gram or horsegram), The bags were tied and kept vertically on wooden racks or suspended inside thatched room where temperature of 26±2°C and relative humidity of 80-95% were maintained. Spawn running occurred for 15-20 days after which the polythene containers were removed and the beds were placed on wooden racks. Pinning was noticed within 3-5 days and the first flush of mushrooms crop was obtained 3-4 days later. Subsequent 2-3 flushes were obtained with a 10 days interval. The total samples, the contents of moisture, crude fibre, ash, total nitrogen (micro-kjeldhal method) and fat were estimated as per the procedures described in AOAC (1975). The content of calcium, phosphorus and iron were estimated as per the methods given by Clarke and Collip (1925), Fisk and Subbarao (1925) and Ramaswamy (1958) respectively.

Table 1: Yield of Oyster mushroom (*Pleurotus*)

		Total yield (kg/bed.)			
Variety	Place of Cultivation	With green gram flour	Without green gram flour	With horse gram flour	Without horse gram flour
White	Haunman-thanagar	1.0–1.71	1.64	–	–
Gray	Haunman-thanagar	2.03	2.00	–	–
Gray	Nandira-juthota	–	–	2.13	2.00

RESULTS AND DISCUSSION

The yield of gray variety of oyster mushroom was more than that of white variety (Table 1). Addition of either green gram or horse gram seed powder, in little quantity along with the spawn, had increased the size and yield of both the varieties of oyster mushrooms. The composition of bio-chemical and mineral constituents of white oyster mushrooms was more when compared with that of gray variety (Table 1).

Low fat content of these edible mushrooms is attractive for the diet therapy where caloric and fat restriction is advised. Gopalari *et. al.,* (1984) compared the nutritive value of mushrooms for protein content as twice that of vegetables like potato and bottle gourd and 4-12 times that of fruits like mango, orange, guava and apple. The composition of Ca, P and Fe in both the varieties of oystar mushrooms was observed to be in sufficient level for human nutrition.

ACKNOWLEDGEMENT

The author is grateful to the Deputy Director, Soil Testing Laboratory, Department of Agriculture, Bapatla (AP) for the liberal laboratory facilities extended to her for the conduct of chemical analysis.

Table 2: Chemical composition of Oyster mushroom (*Pleurotus*)

Variety	Moisture	Protein	Fat	Crude fibre	Ash	Ca	P	Fe
	g 100g				%	mg/100g		
White	90.5	4.9	0.74	1.38	1.46	98	500	17.5
Gray	88.4	3.1	0.65	1.08	1.87	77	476	8.5

LITERATURE CITED

AOAC. 1975. Official methods of analysis of association of official Agricultural Chemists. 12th Ed. Published by the Association, Washington.

Clark, F.P. and Collip, J.B. 1925. J. Biochem. 13. 461-462.

Fiske, C.H. and Subbarao, Y. 1925. J. Biochem. 66, 375-377.

Gopalan, C, Ramasastri, B.V. and Balasubramanian, S.C. 1884. Nutritive value of Indian Foods. 72-92.

Rrmaswamy. W.R. 1958. Biochem. J. 2777-2779.

Singh, N.S. and Zakia Bano. 1977, Ind. Fd. Packer, 30, 18-19.

Shobha Devi, P. and Sarojini, G. 1983, Ind. J. Nutr Dietet. : 0: 222-225.

Subramanyan, T.R. 1986. Mushroom cultivation, ICAR, Bulletin, 8-10.

Vijaya Khader and Narayana, B. P. 1981. Ind. J. Mushroom, 7: 18-25.

❑❑❑❑❑

NUTRITIONAL STUDIES ON PADDY STRAW MUSHROOM

(Volvariella Volvacea)

Indian Journal of Mushrooms, Vol. VII Jan. Dec. 1981, No. 122 PP 19-25

Vijaya Khader and Pandya Nayana B.

(P.G. Deptt. of Home Science, S.P. University, Vallabh Vidyanagar Gujarat-388 120)

Introduction

PADDY STRAW mushroom is a delicious tropical mushroom found growing extensively on paddy straw in India. Overseas, there is only one reference[1] on the proximate composition of *Volvariella volvacea* from Philippines which is however not comprehensive. Zakia Bano *et al*[2] have reported the essential amino acid composition of another variety of paddy straw mushroom (*Volvariella diplasia*). Gopalakrishnan and Pruthi[3] reported proximate composition of paddy straw mushroom. The reported values for moisture, fat, crude protein, carbohydrate, crude fibre and ash are as 92%, 0.21%, 2.94%, 2.19%, 1.15% and 0.99%, respectively. The digestibility coefficient has been reported as 70% and 84% by Munjal and Seth[4] and Thayumanavan and Manickam[5]. Much of the earlier work[6-8] done in India refers to cultivation aspects. Very little published information exists in the literature on the nutritive value of *Volvariella volvacea* which grows extensively in tropical areas.

During the present investigation the cultivation of paddy straw mushroom and the nutritive value of the same has been studied by chemical as well as biological methods. Histological studies were also carried out on liver and kidney.

Material and Methods

Cultivation : Spawn for the cultivation was obtained from International Mushroom Company, Godhra. Tall and stiff variety of paddy straw was obtained from the market, cut into two feet length and made bundles weighing 300-400 grams These bundles were soaked in river water for 10 hours. Drained the excess of water and arranged beds on racks. Edible soyabean cake was obtained from NDDB, Anand.

Bed preparation : Arranged a layer of straw on a rack in a square form of 2′ × 2′ with about 3-4" thickness. Sprinkled two spoons of bengal gram dal powder (besan) on the four sides of the layer keeping a margin of two inches. The spawn was spread on the sprinkled dal powder. Again sprinkled a little dal powder on the spawn. The second layer of the soaked paddy straw was laid at right angles to those of the first layer. Repeated the same for two to three layers. The bed was pressed down firmly by hand and watered immediately and covered it by a transparant polethylene sheet Temperature of the room was maintained between 28-30°C with the help of room heater Water was sprinkled every day to keep straw moist till the mushrooms appeared.

Method of Analysis : Analysis of moisture, fat, potassium and total nitrogen was done by micro-kjeldhal method, crude fibre and ash were carried out by AOAC[9] methods, calcium by Clarke and Collip[10], phosphorous by Fiske and Subbarow[11] and Iron by Ramaswamy method[12]. The calorific value was calculated by multiplying the fat, carbohydrate and protein contents by 9, 4 and 4, respectively.

Experiments

Growth rate of rats and Protein efficiency ratio : Weanling male rats, 6 per group (Wistar strain) were obtained from the Sarabhai Research Centre Stock Colony (Baroda) and these were housed individually in cages with wire mesh at the bottom and fed diets containing dried mushroom and soyabean flour for four weeks. The diets contained in percent, the following ingredients : Oil 10, salt mixture 2[13], vitamin mixture I[11], dried mushroom or edible soyabean cake to provide 10 percent protein in the diet and corn starch to make it to 100 percent. Records of daily food intake and weekly gain in weight of rats were maintained. PER was determined according to the procedure of Chapman *et al*[14].

Collection of Faeces, Liver and Kidney : During the last week of the experiment, the faces were collected for 3 days and dried to constant weight for further analysis. Food intake for these three days were recorded. The rats were sacrified at the end of the period, liver and kidney were collected and weighed immediately. From this 0.5 g of fresh liver and kidney were taken in 10 percent formalin solution for histological study. Remaining liver samples were dried in an oven at 60°C to constant weight for moisture estimation.

Overall digestibility and Apparent digestibility[15] : The following formulae were used for calculation.

Foods & Nutrition Department, College of Home Science, APAU, Saifabad, Hyderabad-500 004.

$$\text{Overall digestibility (\%)} = \frac{\text{Food intake (g)–Faeces excreted (g)}}{\text{Food intake (g)}} \times 100$$

$$\text{Apparent digestibility (\%)} = \frac{\text{Nitrogen intake (g)–Nitrogen excreted in faces (g)}}{\text{Nitrogen intake (g)}} \times 100$$

Histological Studies : The liver and kidney samples were processed[18] and stained with haematoxylin and eosin, and Sudan IV and Haematoxylin.

Results and Discussions

Cultivation : Mushrooms appeared on the 13-15th day of spawning and continued to grow for 2-3 days. A second crop appeared after one week. After that the bed was removed.

Picking : Small buttons of mushrooms are picked twice a day. Care was taken to pick only those heads which have grown fully but before they open up and turn into loose umbrella shapes. The growth of the mushroom was normal without any abnormalities. Since the cultivation was done during the winter season it was rather difficult to maintain the required temperature by room heater. In the present study the paddy straw used for the cultivation is of the common variety which the farmers used for the cattle feed. To provide the extra nutrients for the spawn the red gram dal powder was used. Krishanamohan and Jayarajan[17] have studied the effect of certain organic and inorganic substances to increase the yield of paddy straw mushroom. Tapioca starch flour and horsegram dal gave 88.5 and 67.2 percent increase in yield, respectively over control[17]. The other organic substances to straw beds have been tested and found successful are rice leaves[18] red gram dal powder, bengal gram [19,20], groundnut cake, black gram, green gram and massor dal[21].

Fresh mushrooms after picking, were cleaned properly and then sliced longitudinally and dehydrated under hot air flowing continuously for 20-25 hours at a temperature of 50–55°C. The dried mushrooms were used for proximate analysis and for animal experiments.

Chemical Analysis : It can be concluded from Table 1 that the dried mushroom contain a higher amount of protein and minerals and a low amount of fat and carbohydrate which is the main attraction of many of the Scientists as well as Nutritionists to make use of this particular food in recommending a diet for heart or diabetic patients.

The dried mushroom contain 28% of protein where as literature values ranged between 15 to 44% Gopalakrishnan and Pruthi[3] reported that the *Volvariella volvacea* contain about 37% protein on dry weight. Protein content mainly depends upon the type of species, time of harvesting, storage at which harvesting is completed, the part taken for the estimation and time elapsed between picking and analysis.

Animal Experiment

Growth rate of rats and PER : The results reported in Table–2 indicate that the gain in weight of the rats fed on edible soyabean cake diet is significantly higher. The loss of body weight was observed when the rats were fed on mushroom diet. Three rats fed on mushroom diet died during the 3rd to 4th week of the experimental period. The reason may be that the dried mushroom was used as such without subjecting it to any heat process. The mushroom proteins are generally deficient in methionine, tryptophan and lysine. Besides the amino acid imbalance, the toxic factors present in raw mushroom may also be responsible for growth depression.

Overall digestibility : It appears from the results (Table–3) that the overall digestibility of the dried mushroom and edible soyabean cake differ significantly at 5% and 1% level. Munjal and Seth4 reported similar values.

Apparent digestibility : The values in Table–4 indicate apparent digestibility of the mushroom protein, is very very low as compared with the values obtained with soyabean cake. No data in literature is available.

Mean weight of the body, liver and kidney of rats : There is a marked difference in weight of liver and kidney among the two groups of rats fed on mushroom and soyabean cake (Table–5).

Histological studies on liver and kidney of rats : Normal structure of liver cells has been disturbed, more of vacuoles in structure, swollen appearance and mildnecrosis with cellular infiltration was observed. Nuclei were normal Slight fatty infiltration was observed. Structurally they did not show any marked pathological changes.

Kidney cells are ballooned and nuclei swollen. Collecting tubules were swollen. Medulla and cartex are not very clear. Necrosis of the periphery was observed The cells were degenerated and broken. The pathological changes were more severe in the case of kidney as compared with liver.

Normal histological picture was observed in the case of liver and kidney of rats fed on edible soyabean cake.

Eventhough, chemical data showed that there is more protein in the dried mushroom, it is not biologically available for growth. This can be predicated by the essential amino acids, lysine and methionine which are mainly responsible for growth and which might have been deficient in the mushroom protein. As the raw mushroom powder was used in the diet the toxic factors which are present in the same may possibly be responsible for lowering the nutritive value of mushroom protein.

Acknowledgements

The authors cordially thank Dr. V.C Kotak, Lecturer, Department of Biosciences for extending his cooperation in histological studies Thanks are also due to Dr. D.N. Shah for helping in statistical analysis.

References

1. University of Philippines, Technical Bulletin No. 5, Univ. of Philippines, College of Agri Lagune, Philippines, 1967.
2. Zakia Bano, Srinivasan, K.S. and Singh, N.S. J. Food Sci. Technol., 8, 160, 1971
3. Gopalakrishnan, M and Pruthi, J S. Nutritional quality of Tropical paddy straw mushroom. Indian food Peker, xxxi, 14, 1977.

4. Munjal, R.L., Seth P.K , and C L. Jandaik. Cultivations of *Agaricus bisporus* in special bulletin of the Indian Mushroom Growers' Association. Mushroom cultivation in India, 1, 1977.
5. Thayumanavan, B and Manickam, A. The Indian J. Nutr. Diet. 17, 140, 1980.
6. College of Agri. Cultivation of tropical mushrooms; Bull. Tamilnadu Agri. University, Coimbatore, Tamilnadu, 1968.
7. Agri. Information Service, Himachal Pradesh, Simla. How to grow Mushroom- Bull. Agri. inf serv. 1, 1972.
8. Zakia Bano and Srivastava, H.C Food Sci. 12, 363, 1962.
9. AOAC Official Methods of Analysis of Association of Official Agricultural Chemists, 12th ed. Published by the Association, Washington, 1975.
10. Clark, F.P. and Collip, J.B. Journal of Biochem. 13, 461, 1925.
11 Fiske, C.H and Subbarow, Y J. Biochem. 66, 375, 1925.
12. Ramaswamy, W.R. Biochem. J. 53, 277, 1953.
13. Hubbell, R.B., Mundel, L.B. and Wakeman, A.J J. Nutr. 14, 273, 1937.
14. Chapman. D.G., Castillo, H and Campbell, J.A Can. J. Biochem. Physiol, 37, 679, 1959.
15. Vijaya Khader, Ind J. Nutr. Diet. 16, 316, 1979.
16. Haw, A W. Histology, J.B. Lippincott Co., London, 9, 1950
17. Krishanamohan, G. and Jayarajan, R. Mush Sci. 1, 425, 1976.
18. Chang, S.T. The Chinese mushroom. The Chinese University, Hong Kong, 1972.
19. Asthana, R.P. Mag. Agric. Coll. Nagpur. 22, 26, 1947.
20. Vasudeva, R S Sci. Rept. Ind. Agril. Res Inst., New Delhi, 87, 1955.
21. Gupta, G.K., Bajaj, B.S. and Suryanarayana, D. Iud. Phytopath. 23, 615, 1970.
22. Cartter, C.D. and Hopper, T.H. U.S.D.A. Technical Bulletin, 787, 1942.

Table–1 Proximate composition of dried mushroom and edible soyabean cake.

Constituents per 100 g	Dried Mushroom	Edible Soyabean cake*
Moisture (g)	6.3	8 0
Protein (g) (Nitrogen × 6.25)	28.0	45.7
Fat(g)	3.8	–
Crude fibre (g)	4 9	2.5
Carbohydrate (g)(by difference)	49.7	30.0
Ash (g)	13.7	6.2
Calcium (mg)	34.6	300.0
Phosphorus (mg)	911.1	600.0
Iron(mg)	16.5	14.0
Potassium	1.7	—
Calorific value	336.0	302.8

*Cartter and Hopper (22).

Table–2 PER and growth rate of rats fed on mushroom and edible soyabean cake (Experimental period 4 weeks, 6 male rats per group, Protein level in diet 10%).

Group No.	Diet	Protein in diet (g) (N × 6.25) on dry wt. basis	Food intake (g)	Protein intake (g)	Gain or loss in wt. (g)	PER
1	Mushroom	10.0	121.0	12.1	–7.5	*
II	Edible soyabean cake	10.0	396.3	39.6	102.0	2.5

* All the rats lost weight, PER was not calculated.

Table–3 Overall digestibility of mushroom and edible soyabean cake (Duration of experiment one week, 6 male rats per group, protein level in diet 10%)

Group No.	Diet	Food intake (g)	Faecal weight (g)	Overall digestibility	't' Value
I	Mushroom	12.5	1.59	87%*	2.262 (5%)**
II	Edible soyabean cake	50.5	3.34	93%	3.252 (1%)

*Values for five rats. Statistical analysis by student 't' test.

**Significant at 5% and 1% level.

Table–4 Apparent digestibility of mushroom and edible soyabean cake (Duration of experiment, one week, 6 male rats per group, protein level in diet 10%).

Group No.	Diet	Nitrogen intake (g)	Faecal nitrogen (mg)	Apparent digestibility
I	Mushroom	140	0.88	37%*
II	Edible soyabean cake	8.97	1.41	84%

* Values of five rats.

Table–5 Mean weight of body, liver and kidney of rats fed on mushroom and edible soyabean cake (Duration of experiment 4 weeks, 6 male rats per gourp, protein level of diet 10%).

Group	Diet	Mean body weight		Wt. of liver on fresh basis (g)	Wt. of liver on dry wt. basis (g)	Moisture %	Wt. of kidney fresh wt. basis (g)
		Initial (g)	Final (g)				
1	Mushroom	42.17	35.25	1.30*	0.47	93.3	0.63*
II	Edible soyabean cake	42.17	144.17	5. 97	0.80	97.7	1.20

*Average for four rats.

❑❑❑❑❑

ACCEPTABILITY STUDIES ON WEANING FOODS AND PICKLES

(PREPARED OUT OF PADDY STRAW MUSHROOM (Volvariella Volvacea) AND THE KEEPING QUALITY OF THE SAME)

Indian Journal of Mushrooms, VolII, Jan.-Dec. 1981, No. 1 & 2 PP 32-36

Vijaya Khader and Pandya Nayana B.

(P.G. Deptt. of Home Science, S.P. University, Vallabh Vidyanagar Gujarat-388 120)

Introduction

MUSHROOMS are the noval source of food which require popularization among the local community. The acceptability of the preparation is essential for the consumers to accept the same. Fresh mushrooms are highly perishable in nature and cannot be stored for more than 24 hours at room temperature and in the peak period of harvesting they are much in excess than demand. To prevent such gluts in the fresh market it is necessary to preserve this product by different methods Canning of mushroom is considered to stabilize the industry and make the product available beyond growing season[1].

Preservation of mushroom by canning[2] and dehydration[3] and pickling is in practice for quite some time. No studies have been carried out on the use of mushrooms in the preparation of weaning foods which play a vital role in the developing countries when two thirds of population is suffering from malnutrition The present report covers the acceptability studies on pickles as well as weaning foods prepared out of mushrooms and keeping quality of the same.

*Foods and Nutrition Department, College of Home Science, APAU, Saifabad, Hyderabad-500 004.

Materials and Methods

Materials : Paddy straw mushrooms grown in the Laboratory, P. G Department of Home Science, S.P. University were boiled for five minutes and dehydrated at 60°C, was ground to a fine powder before incorporation in the weaning foods. Fresh mushrooms were used for pickle preparation.

Weaning foods : Attempts were made to prepare weaning foods using various combinations, is given below.

Weaning Foods

Weaning food I

Puffed rice powder	–	20g
Whole Bengal gram (germinated, dried roasted).	–	20g
Mushroom powder	–	20g
Jaggery	–	10g
Cardamom	–	One

Preparation : Above ingredients were mixed thoroughly and packed in a clean dry glass bottle with cover.

Weaning food II

(*a*)				(*b*)
Rice powder	–	40g	–	Kodra powder
Red gram dal powder	–	10g	–	Green gram dal powder.
Bengal gram dal powder	–	10g	–	Bengal gram dal powder
Mushroom powder	–	40g	–	Mushroom powder

Preparation : The cereal or millet, pulse and mushroom powders were mixed and prepared batter and fermented overnight. Jaggery 10g, salt to taste and pinch of soda were added and mixed well and baked at 150°C for 10-15 minutes. Allowed to cool at room temperature dried well and powdered, packed in a clean dry glass bottle with cover.

Mushroom Pickle as per Singh and Zakia Bano[4] : Fresh mushroom were chopped with stainless steel knife, washed in clean water twice and squeesed gently to remove excess water lodged in between the gills.

Ingredients	*Amount*
Mushrooms chopped	125 g
Cumin seed powder	1/2 tsp
Fenugreek seed powder	1/2 tsp
Coriander seed powder	1/2 tsp
Turmeric powder	1/2 tsp
Mustard seed powder	1/2 tsp
Green chillies	5 Nos
Vinegar (Commercial)	45 ml
Sesame oil	45 ml
Salt	to taste

Method : The prepared mushroom was cooked in 30 ml sesame oil with salt in a covered stainless pan on simmering flame for 10 to 15 minutes and kept aside. All the spice powders were mixed together. Green chillies were slitted longitudinally and slightly fried in about 15 ml of oil. Spice mixture was added and mixed well in oil. Then the cooked mushrooms were added, heated to boiling and filled in bottle after cooling Caps were screwed on the bottle and bottles were kept at room temperature.

Mushroom pickle as per Kanpur Women's Mushroom Club*

Ingredients	*Amount*
Mushroom (fresh)	125 g
Mustard seed powder	1/2 tsp
Turmeric powder	1/2 tsp
Chilli powder	1/2 tsp
Salt	1/2 tsp
Asdfoeteda	Pinch
Sesame oil	30 ml

Method : Boiled the mushroom for 10 minutes and drained the water and dried in sun light for 2-3 hours. Oil, spices and mushrooms were mixed and kept in a jar and the jar was kept in sunlight for 2-3 hours and then kept at room temperature.

Chemical analysis : Analysis of moisture, fat and potassium, total nitrogen by micro-kjeldahl method, crude fibre and ash were carried out by AOAC methods[5], calcium by Clark and Collip[6], phosphorus by Fisk and Subbarow[7] and Iron by Ramaswamy method[8]. The calorific value was calculated by multiplying the fat, carbohydrate and protein contents by 9, 4 and 4, respectively.

Acceptability Studies

Selection of Panel members and score card : The taste panel consists of 15 members from the P.G. Department of Home Science and ranking test[9] was used to find out the acceptability of the preparations.

Shelf-life of the weaning Foods and Pickles : The weaning foods as well as pickles were kept at room temperature for a period of 5 months and regular observations were carried out to see the shelf-life of the preparations. When the spoilage of the pickle was observed by naked eye, the microbiological test (Simple staining and gram staining)[10] was carried out to detect the type of microorganism.

Results and Discussion

Consumption of mushroom is not very common with the local community where people are so religious minded and believe very strongly that the mushrooms are "bilodine top" (cat tail) and are hesitent to prepare it in the house.

The chemical composition of weaning food I, II (a) and II (b), dried mushroom are given in table I. Weaning food I contain higher amounts of fat as compared with other weaning foods. This may be due to the reason that in the preparation of weaning food

---

*Personal Communication.

I, 10% of fat was used where as in preparation of other weaning foods no fat was used. The present results indicate that the dried mushroom powder contain 3.8% of fat which is comparable with the results of Gopalakrishanan and Pruthi[11]. Protein content is slightly higher in the case of weaning food I. The chemical analysis indicate the weaning food I is better as compared with other two weaning foods as far as protein, fat, calcium and calorific value is concerned.

Acceptability studies indicate that the scores for taste, flavour and overall acceptability for both the pickles are markedly different from each other. The pickle prepared as per Singh and Zakia Bano[4] is found to be more acceptable to all the taste panels as compared with the pickls prepared as per Kanpur Women's Mushroom Club The reason may be due to the raw oil and masala which were used in second pickle preparation.

All the three weaning foods weaning food I and weaning food II (a) and (b) are accepted well with narrow chance in Score. This indicates that all the three weaning foods are quite acceptable without any , emark

Microbiological Studies : The prepared pickles from the fresh mushrooms were kept at room temperature to see the keeping quality. After five days the second pickle got spoiled. This may be due to the use of raw oil and the fresh mushroom boiled in water, kept in sunlight for 2-3 hours instead of 2-3 days. The culture was prepared from the spoiled pickle for microbiological studies. Gram staining technique was used to identify the microorganisms. The results showed the presence of streptococci and staphylococci, which are mainly responsible for spoilage of pickle. First pickle after five months showed some viscosity that might be due to the green chillies added to it which contain some amount of moisture. The gravy looked more viscous may be due to the formation of emulsion with moisture and oil.

The keeping quality of weaning foods was quite satisfactory up to a period of five months as indicated by the acceptability studies. The microbiological observations showed no drastic changes in the weaning foods at the end of fifth month.

Summary and Conclusion

The chemical composition of the weaning food I (puffed rice + Bengal gram + Mushroom) contain more of fat, calcium, iron and calorific value as compared with weaning II (a) (Rice + Red gram dal-+-Mushroom) and II b (Kodra + Green gram dal + Mushroom). The chemical analysis indicate that the dried mushroom is a good source of protein as well as in minerals.

All the three weaning foods prepared in the laboratory were well accepted by all the panel members. The pickle prepared as per Singh and Zakia Bano[4] was preferred better.

The pickle prepared as per Kanpur Women's Mushroom Club got spoiled within five days while first pickle stayed for about five months without fungal growth.

The microbiological observation of spoiled pickle showed the presence of streptococci and staphylococci.

Acknowledgements

The authors are thankful to Dr. S.S. Sail, Head of the Department of Home Science for providing the facilities.

References

1. Dany, R.L. and Singh, R.P. : Preservation of mashrooms. Mushroom Science 1 : 215-222 (1976).
2. Food Technology the world over, Ed. by Peterson, M.S. and Tressler, D.K. The AVI Publ. Co , Inc., West Port, Connecticut, 1965, vol. 11, p. 299.
3. Food Dehydration, Ed. by Van Arsdel, W.B. and Copley, the AVI Publi. Co.-, Inc., West Port, Connecticut, 1964, Vol. II, p. 257.
4. Singh, N.S. and Zakia Bano. Indian Food Packer Vol. XXXI No. 5, 18-19 (1977).
5. AOAC Official methods of Analysis of Association of Official Agricultural Chemists, 12th ed. Published by the Association, Washington (1975).
6. Clarke, F.P. and Collip, J.B. Journal of Biochem. 13, 461 (1925).
7. Fiske, C.H and Subbarow, Y.J. Biochem. 66, 375 (1925).
8. Ramaswamy, W R. Biochem. J 53, 227 (1953).
9. Swaminathan, M- Sensory methods of analysis of foods. Essentials of Foods and Nutrition. 2, 164-173 (1974).
10. Mograw, Manual of Microbiological methods. Committee on bacteriological technique. New York (1961).
11. Gopalakrishnan, M. and Pruthi, J.S. Indian Food Packer Vol. XXXI. 1977. 14-17.

Table–1 Chemical composition of dried mushroom and weaning foods.

Constituents (Value per 100 g)	Weaning foods			Dried Mushroom
	I	II(a)	II (b)	
Moisture (g)	11.7	4.8	4.9	6.3
Protein (g) (Nitrogen X 6.25)	27.0	15.8	16.7	28.0
Fat (g)	12.9	3.8	4.6	3-9
Crude fiber (g)	1.2	1.5	1.8	4.9
Carbohydrate (g) (by difference)	63.7	71.0	69.2	49.7
Ash(g)	5.2	7.9	7.7	13.7
Calcium (mg)	72.0	58.0	66.0	34.6
Phosphorous (mg)	165.0	228.0	311.0	911.1
Iron(mg)	2.1	1.6	1.7	16.5
Potassium (g)	1.4	1.6	1.6	1.7
Calorific value	478.9	381.4	385.0	336.0

Weaning Food 1 Puffed rice (20 g)+ Bengal gram (20 g)+ Mushroom (20 g).
Weaning Food II (a) Rice (40 g) + Red gram dal (10 g) + Mushroom (40 g).
Weaning Food II (b) Kodra (40 g) + Green gram dal (10 g) + Mushroom (40 g).

❑❑❑❑❑

IMPACT OF TRAINING ON CULTIVATION OF MUSHROOMS TO THE UNEMPLOYED

Rural India, June-July 1993, PP 123-125

Vijaya Khader

Dr. Vijayakhader, Prof and Head, Dept. of Foods and Nutrition, College of Home Science, Head, Dept. of Foods and Nutrition, College of Home Science, Bapatla-522 101, Guntur District—(A. P.)

Successful Cultivation of Mushrooms on a commercial basis calls for an exposure of the future growers to mastery of technology., So that it will provide avenue for a profltable self employment. Disguished unemployment has became a characteristic factor of an agricultural sector. Growing of Mushrooms can go a long way in the efficient uiitization of agricultural waste (Sutherben, 1987). Nutritive value of Volvariella Valvacea and Pleurotus was studied by Vijayakhader and Nayana (1981) and Vijayakhader (1988) Mushroom culture is an excellant subsidiary occupation suited for our country. To promote self employment avenues for rural families, training programmes were conducted in Mushroom Cultivation.

The Andhra Pradesh Agricultural University has made an attempt to train illiterate unemployed men and women on various aspects of edible mushroom cultivation includes spawn production, nearly 300 people were trained on the cultivation of Pleurotus Species at the College of Home Science, Bapatla in different batches from different districts such as Guntur, Krishna Visakhapatnam. Nalgonda, Prakasam Nellore,

Rangareddy West Godavari, East Godavari. A few persons from Orissa, Madras also underwent training

The duration of the training was 3 days. After allowing a reasonable period of putting the technology into practice, a study was undertaken with the following objectives;

1. To find out the extent of adoption of mushroom cultivation by the trained women and men.
2. To find out the usefulness of the training course adequacy of the course contents and constraints in adopting at different stages of mushroom growing as pereccived by the trained personal.

Materials and Methods

A structured questionnaire consisting of 20 questions on various aspects of the cultivation of edible mushrooms was prepared and mailed to 300 trained persons. The age of the trained ranged from 15-40 and most of them are between 20–30 years. Out of 300 respondents 200 (6-6.7%) returned the questionnaire. Frequency and percentage were worked out for interpretation of results (Rahiman and Nehru 1983).

Results

Existence of any business system depends upon demand for its products It has always to be kept in view that customr is the nucleus of all the business actively which may be agricultural, industrial or of any other kind. Though the trained persons were engaged in homestead farming, 30.2 per cent had busines as main occupation and remaining 28 per cent were unemployed.

Motivatioual Factors to Join the Training Programme

Six statement which could be the possible motivational factors to join the training on mushroom growing, were given and asked them to indicate the most significant one as perceived by them. The results are given Table-1.

Table No.–1 Motivational Factors for Joining the Training

S. No.	Motivational Factors	Frequency	Percentage
1.	To adopt as an occupation	80	40
2.	To grow mushroom for self employment	50	25
3.	Just to acquire knowledge	10	5
4.	To get practical training	10	5
5.	For the sake of a certificate	30	15
6.	To establish a linkage with other countries	20	10
	Total	**200**	**100**

Usefulness of the Training :

Putting the knowledge gained into . practice is an indication of the usefulness of the training. The responses of the trained youth about the usefulness of the training were encouraging. Majority (90%) of the respondents stated that because of the training, the, could distinguish between edible and poisonous mushrooms to a certain extent, and 10; percent could identify very easily. When asked about the extent of utilization of knowledge in different stages of mushroom cultivation the responses obtained are presented **Table-2.**

Table No. 2 Extent of Utilization of Knowledge Gained Through Training,

S. No.	Stages of Mushroom Growing	Frequency	Percentage
1.	Preparation of Pure Culture	10	5
2.	Preparation of Spawn (Bottle to Bottle Transfer)	80	40
3.	Lay out of Beds and Growing Mushrooms	100	50
4.	Preparation of Spawn and Growing Mushrooms (2831)	10	5
	Total	**200**	**100**

Course Contents and Duration of Training

In respect of course contents, 80 (40 per cent) of the trained persons were of opinion to increase the practical where as 100 (50 percent) expressed the view that the proportion of theory to practicals was adequate. The ramaining 20 persons (10 percent) wanted increase the theory classes. The present duration of three days was found adequate majority (85 per cent) while only 15 percent to increase the duration from three to six.

Constraints in the Cultivation of Mushrooms.

Based on prelimainary discussions held with some Mushroom growers, six condstraints were listed out and the respondents were asked to rank them from 1 to 6 accordance to their importance, as perceived by them. The mean score for each constraint was worked out and ranked. The resulted are presented in Table–3.

Table No. 3. Constraints in the Cultivation of Mushrooms

S. No.	Constrants	Ranks
1.	Difficulty in mother spawn preparation	I
2.	Problem of Pests and Diseases	VI
3.	Lack of Financial Assistance	V
4.	Mushroom Growing is not Profitable	IV
5.	Lack of Market for Mushroom	III
6.	Lack of awareness on the part of people about 'preservation' aspects of mushroom	II

Discussions

Though the training was priliminary ment for unemployed youth, majority of (72%) them were engaged in farming or business. Only 28% were totally unemployed. This points to the enthusiasm among the youth in getting themselves engaged in any gainful occupation. It was also encouraging to note that 40 percent of the participants joined the training course with a view to adopt mushroom as an occuption, and 95 percent of them put to use the knowledge gained either in preparation of spawn (bottle to bottle transfer), in laying out mushroom beds, in spawn preparation and laying out beds for growing mushrooms. But lack of adequate knowledge of mother spawn preparation, lack of awareness on the part of people about preservation aspects of mushrooms and also lack of adequate market for the produce were the most important constraints in mushroom growing followed by lack of financial assistance.

Creating awareness among the public about the nutritive value and therapeutic use of mushroom cultivation through mass media is suggested by trained youth would perhaps go a long way in finding a market for the produce. Simultaneously steps need be initiated to make avaiable good quality of spawn through recognise universities and financial assistance through nationalised banks to the mushroom cultivation.

The training programme, no doubt, could enthuse the illiterate youth in mushroom growing. Constraint follow-up on the part of the trainers and encouragement by the Governmental agencies would help to perpetuate this enthusiasm.

❑❑❑❑❑

I MAP TODAY

Indian Medicinal and Aromatic Plants Today

Infoconcepts India Inc. July-August, 2004

India's Ist Bi-Monthly Megazine Exclusively for MAP Sector

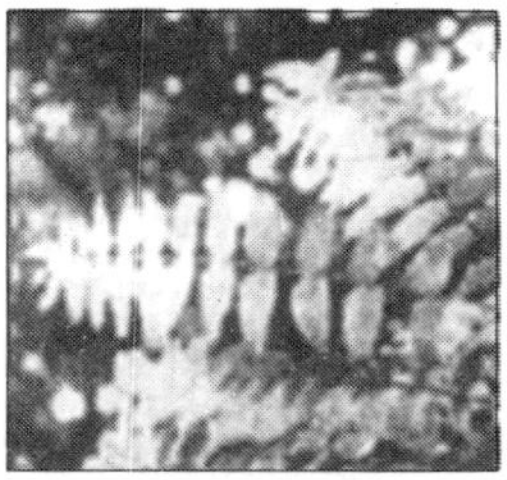

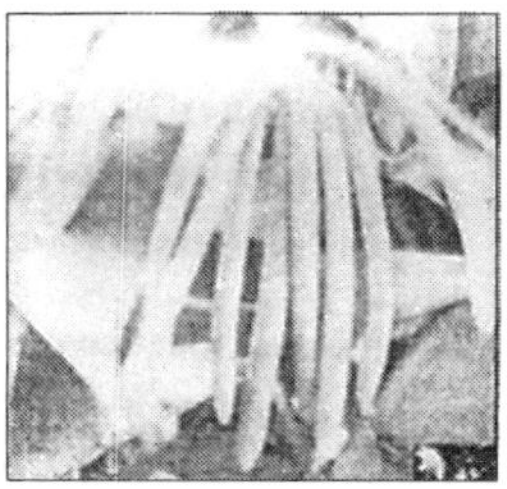

SUITABLE TECHNOLOGY FOR CULTIVATION OF MUSHROOM IN SMALL FARMS

During the recent years, the commercial cultivation of mushrooms in India is gaining momentum as one of the significant export-oriented agro-industrial enterprises due to the advantages of rural employment generation and gainful utilization of natural and farmer resources. Mushroom cultivation is emerging in India as a promising agro-based. land-dependent enterprise due to flexibility in its operation. About 2250 man days of employment are generated in production of 9450 kg of white button mushroom. About 1000 man days are required to produce 4800 kg of oyster mushroom, where as 50 man days to produce 360 kg of paddy straw mushrooms.

Suitable technologies for socio-economic upliftment.

Oyster Mushroom Cultivation - Self Employment

Mushroom cultivation has a special relevance to India because straw and plant residues are abundantly available with farmers and environment is conducive to the cultivation of mushroom. Hence the future of this industry is bright. The awareness for mushrooms cultivation among the general public has been created. Mushroom growing as a cottage industry is quite valid for India grower's due its low capital investment and high yields obtain even under controlled rural conditions. Mushrooms are good supplement for protein lacking diets and can be easily cultivated indoors and marketed profitabily.

Material needs

(*a*) Paddy straw
(*b*) Polythene bags
(*c*) Vessels
(*d*) Spawn
(*e*) Pressure cooker
(*f*) Fuel

CULTIVATION PROCEDURE

1. Raw Material (Paddy straw/Paddy husk/any Cellulose Material)
2. Cut into 3.0-5.0 m pieces
3. Soak in water (overnight or 12-16)
4. Drain excess water
5. Sterilize the straw bits in boiling water (15-30 mins)
6. Fill the polythene bags 60 cm length and 30 cm width with sterilized straw bits (5 cm ht.)
7. Sprinkle handful of spawn (seed of mushrooms) over the straw layer
8. Again cover with straw bits to another 5 cm ht.
9. Repeat alternatively spreading the spawn and straw bits till the bag gets filled and Tie with thread (known as bed)
10. Arrange the beds in shelves leaving a space of six inches between the beds
11. After 12-15 days remove the polythene cover.
12. Sprinkle water twice a day to keep the the bed moist.

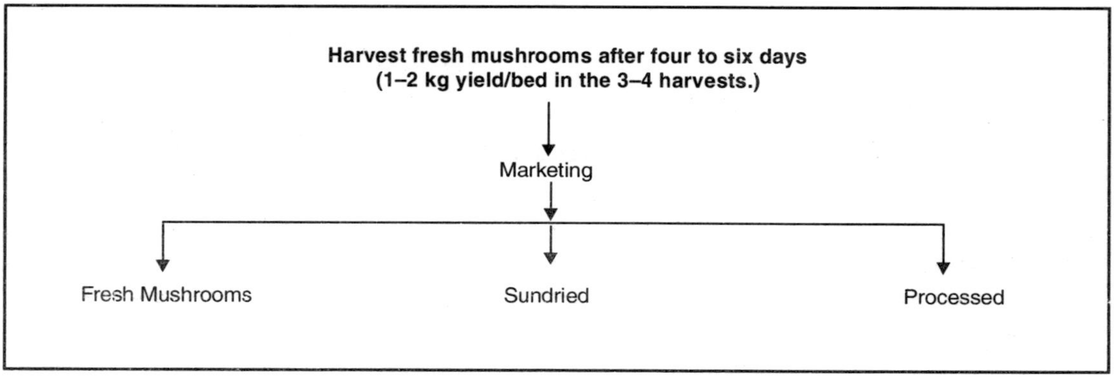

Suitable Technology for Cultivation of Mushroom in Small Farms

Market potential

The studies reveal that mushroom cultivation can be taken up by rural women to have supplementary income as the technology is very simple, require less investment and less space.

It is also evident from various studies spawn multiplication can be done by women as co-operative venture and mushroom cultivation can be taken at household level as an income generating activity.

Estimated cost of cultivation per one Mushroom bed

Expenditure	Cost (Rs.)
Cost of 1500 gm of rice straw at the rate of 20 paise/kg	0.30
Cost of 1/2 bottle of mushroom spawn at the rate of Rs.6/bottte i.e., 300 gm of spawn	3.00
Cost of bamboo or wooden rak	10.00
Cost of polythene bag and miscellaneous expenditure	2.00
TOTAL	15.30

The net income varies and depends on the sale price of the spawn. The income obtained after deducting the recurring cost is Rs. 44.70/-. In the rural areas a woman can easily manage 4-10 beds depending on the space available, helping them to earn Rs. 180/- to Rs. 450/- per month as a supplementary income.

INCOME

Cost of 2 kg mushrooms (Rs. 30/kg) 60 rupees. (The market price varies between Rs.20-40/kg from place to place

Net Income : Rs. 60 – Rs. 15.30 = Rs. 44.70/-

❑❑❑❑❑

RURAL WOMEN AS ENTREPRENEURS IN MUSHROOM CULTIVATION

Indian Forming March 1994, PP 18-21

EVERY woman is an entrepreneur as she manages, organises and assures responsibility for running her house. It has been increasingly realised that women possess entrepreneural talent which can be harnessed to create employment opportunities. The main concern now is to develop women's skills in agricultural and other marketable goods and then make them partners in production and beneficiaries.

The participation of women in economic actvities in India is only 10%, where as it is 40% in developed countries. Increasing food production to meet our country's needs is essential. So there is an urgent need to identify, cultivate and popularise new foods which can not only augment the food deficits but also are nutritious and easily available.

At present it is essential for rural women to go for additional income using simple techonologies. Rural women can cultivate mushroom at home. Various studies indicate that mushroom cultivation can be taken up by rural women to have supplementary income for their family as the techonology is very simple, requires less investment and less space.

An attempt was therefore made in the present investigation to see the effect of mushroom cultivation towards the development of rural women entrepreneurship. In the present study the spawn multiplication and oyster mushroom cultivation at home level was taken up with the main objective to find out the feasibility of spawn multiplication, mushroom cultivation as an income-generating activity at household level and to assess the women's attitude towards mushroom cultivation at home level.

Table–1. Inoculation of sorghum grain with mother spawn

Method	Good growth of mycelium	Contaminated mycelium
Using spirit lamp in sterlized environment	50 Bottles	Nil
Under ultraviolet in the inoculation chamber	50 Bottles	Nil

Table–2. Quantum yield of mushroom

Mushroom yield in grams	Respondents	Percentage
Less than 1500	10	25
1501-2000	20	50
2001 -and above	10	25
Total	40	100

Table–3. Cost of spawn multiplication in 100 bottles

Expenditure	No. of Units	Amount
Non-Recurring		
Autoclave (Electrical coil fixed drum)	1	5,000
Aluminium vessels with lid	2	200
Spirit lamp	1	25
Kerosene Stove	1	100
Recurring		
Mother Spawn	1 Bottle	6
Sorghum	25kgs	100
Calcium carbonate	500gms	15
Empty glucose saline bottles	100	50
Misc. Charges		50
Labour Charges		30
Non Absorbent Cotton	2 Bundles	50
Total		**301.00**

Table–4. Estimated cost of cultivation

Expenditure	Cost
Cost of 1500 gm of rice straw at the rate of 20 paise/kg	0.30
Cost of 1/2 bottle of mushroom spawn at the rate of Rs 6 / bottle i.e. 300gm. of spawn	3.00
Cost of bamboo or wooden rack	10.00
Cost of polythene bag and miscelaneous expenditure	2.00
Total	**15.30**

Income

Cost of 2 kg mushrooms from each bed in 4 pickings at the rate of Rs 30/kg comes out to be Rs 60 (The market price varies between 20 to 40/kg from place to place)

Net Income is Rs 60 Rs 15.30 = Rs 44.70

Mushroom-cultivation is an income-generating activity that can be taken up at household level. Here women are seen preparing mushroom bed (above) and arranging prepared bed on a rack.

Polythene bags containing spawn in layers is placed at room temperature for 15 days.

Methodology

Selection of Samples

Among the mushroom growers 20 respondents from low income and 20 respondents from middle income were selected randomly from 3 places around Bapatla. Selected respondents were given training at the College of Home Science, Bapatla.

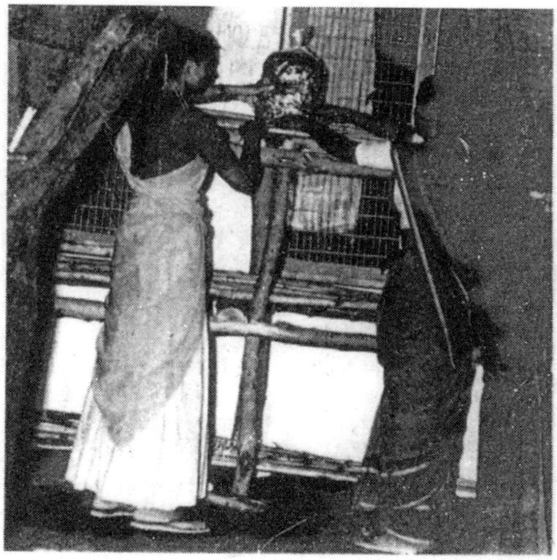

Spawn Multiplication in the Lab

As in any agricultural crop mushroom cultivation also requires good seed naterial. This seed material is called spawn. 100 sterile bottles (500 ml glucose/saline bottles) were filled with half cooked sorghum mixed with calcium carbonate (20g/kg of sorghum) upto three-fourth of the bottle. These bottles were then sterilized in an autoclave/pressure cooker for 90 minutes at 20 lbs pressure. Out of the 100 sterile bottles 50 were inoculated with mother spawn near the spirit lamp in sterile environment and other 50 were inoculated with mother spawn under the ultraviolet lamp in the inoculation chamber.

Field Experiment for Mushroom Cultivation

The selected respondents were shown the various steps in mushroom cultivation (*Pleurotus sajor-caju and Pleurotus citrinopileatus*) through demonstration. They were given spawn bottles for cultivation of mushrooms in their own houses. One and a half kg of substrate were filled into polythene bags of 60 × 30 cms. Spawning was done in 5 layers. These bags were incubated at room temperature (28°C for a period of 15 days. At the end of 15 days beds were removed from polythene bags and kept on the rack for fruiting. Water was given twice a day by spray pump and humidity was maintained between 80-90%by sprinkling water on the gunny bags tied in the room. Harvesting was done before spore shedding. Fresh weight of mushroom was recorded immediately after harverting. Small portion of the straw all over the surface of bed was scrapped in order to harvest the second crop. Three flushes were obtained in a period of 45 days. Periodical visits were made by the investigator as a follow up for the study. First harvest was done after 18-20 days after spawing. Three more harvests were done at 5-7 days intervals.The yields from four harvests were noted down.

Harvesting of crop of mushroom is done before spore shedding. The flushes can be obtained in a period of 45 days. A yisld of 1.5 to 2 kg of mushrooms is obtained with 300 gms of spawn.

Results

There was no problem in using the pressure cooker for sterilization except that it holds less number of bottles.

Table l shows that all the 100 bottles were having good growth of mycelium revealing that inoculation of bottles with mother spawn could be done by using spirit lamp also.

Table 2 shows that nearly 50% of the respondents were able to get an yield of above 1.5 to 2 kg of mushrooms.

During four harvests with 300 gms of spawn which is equal to the yield obtained by the research studies and large scale mushroom industries.

Fresh weight of mushroom is recorded immediately after harvesting

Economics of Mushroom Cultivation

Economics of mushroom cultivation mainly depends on the equipments, materials used, their initial cost and yield performance.

Table 3 shows that the intial non-recuring expenditure was found to be high for the women from low income group. The cost could be reduced if the women could share the non-recurring expenditure collectively. The cost of spawn for one bottle in the market is Rs 6 from the voluntary organization and agricultural universities where as from the public sector it is Rs 10.

The expenditure for preparing the bed is Rs 15-10 for bed which includes the cost of straw, polythene bags, labour charges and spawn (Table 4).

The net income varies and depends on the sale price of the spawn prepared by the entrepreneur. The cost of cultivation is much less thereby increasing the profit. The income obtained after deducting the recurring cost was Rs 45.

In the rural areas a woman can easily manage 4-10 beds depending on the space available, helping them to earn Rs 180 to Rs 450 per month.

After proper training and demonstration the respondents changed their attitude towards mushroom cultivation and have shown very positive attitude towards its cultivation.

Spawn multiplication can be done by women as a co-operative venture and mushroom cultivation can be undertaken at household level as an income-generating activity.

Vijayakhader is Professor and Principal, College of Home Science, Andhra Pradesh Agriculture University Bapatala-522101

❑❑❑❑❑

RURAL WOMEN-NUTRITIONAL SECURITY

Indian Farming Oct 1998, PP 7, 61-63.

The Post-Independence period has seen many positive steps to improve the socio-economic status of women. The Shramshakti Report 1988, for the first time, underlined the critical contribution of the marginalized poor women to the growth of the formal economy in both rural and urban areas.

Participation of rural women in economy growth

> Achieving a well-fed world means improving what people eat, the quantity,. quality and variety. For the hungry, it means increasing their access to food so that they can eat enough. Women plays a key role in food and nutrition security: at family level, community level and national level.

WOMEN IN UNORGANIZED SECTOR

The participation rate of women rose from 23.1% to 27.2% in the rural areas and from 8.3% to 9.7% in urban areas. The Human Development Report 1990, published by the UNDP found that the women labour spent 11.3 hr, whereas the male labour spends only 8.7 hr. The employment of women in the unorganized sector is divided into 9 sectors, viz. *agriculture, dairy, small animal husbandry, fisheries, social and agro-forestery, khadi and village industry, handlooms, handicrafts* and *sericulture.* Many enterprises, have shifted from the organized to the unorganized sector. Among these industries are *coir, cashew, bidi, jute, brick* etc., with enormous market potential. The *garment* and *construction industries* also offer enormous opportunities. Coir industry of Kerala employees about 0.5 million women. The number of women members in India's milk co-operatives has gone up to 1.25 million in recent years.

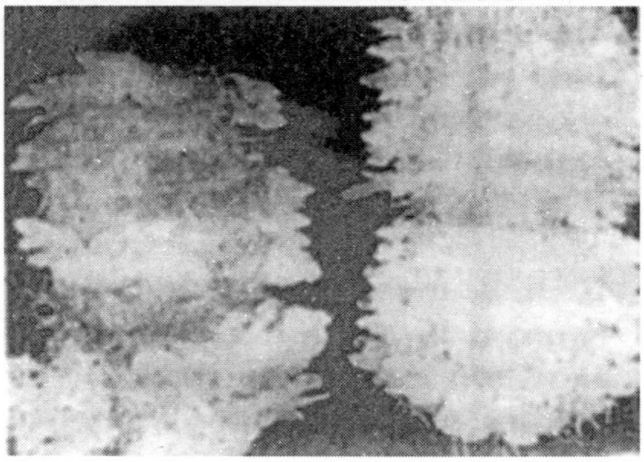

Rural women can cultivate mushroom at home

Sterilization of Sorghum in Pressure Cooker
(Women emoloyment increases their status and income)

In the agricultural sector, mechanization and modernization have displaced women from their traditional jobs to some extent. However, in rice cultivation, there is heavy dependence on women. The percentage of women employees in the public sector, which was 12.84 in 1992 rose to 13.36 in 1998. Around 73% of women entrepreneur launched their projects in the eighties and this trend is steadily continuing in the nineties.

WOMEN ENTREFRENEURSHIP IN INDIA

In India, village and small scale industries are treated as an integral and continuing element in the economic structure. Employment for women is essential not only to increase their status but also to supplement their family income. Women have an important role to play in synthesizing social progress with economic growth of the developing countries. It is an accepted fact that *the role of women in national development is a crucial one.*

Woman feels pleasure in providing food to her family

Backyard animals

Backyard animals are handy sources of income and milk, meat and eggs for the family. Cattle, buffaloes, sheep and goats thrive on marginal land and convert crop residues and byproducts into milk and meat. In India, a well-organized dairy co-operative movement has tapped milk produced by farmers keeping 1 or 2 cows or buffaloes. As a result, over 5 million farm families in 49 000 village co-operatives sell on an average about 8 million litres of milk every day, retaining 25% for their own consumption.

FOOD SECURITY FOR ALL

Food security means all people at all times have access to the food they need for a healthy and active life. It arises from stable food supplies that are both physically and economically accessible to all.

Food security can only be achieved through broadly-based actions that address many factors determining people's ability to obtain the food they need. Agriculture, forestry and fisheries do not simply produce crops and commodities; they provide livelihoods and improve nutrition too. Their products create employment both directly, on the land and seas, and indirectly in processing and related industries. Production should be geared to deliver a wide range of foods.

Economic development in rural areas increases household income. Many farming households obtain as much as two-thirds of their income from off-farm, non-agricultural sources.

Studies have shown that, after food processing, small-scale forest based enterprises provide a major source of off-farm income in many countries. This income is particularly important for the poorest, the landless and rural women.

LINKAGES BETWEEN FORESTRY, FOOD SECURITY AND WOMEN'S LABOUR

The role of women vary in different societies in meeting their household's basic food needs. While men have greater access to cash economy and often generate cash as their primary activity, women's activities revolve more and more around subsistence needs of the household, particularly *food production* and *child care. Women also contribute to household cash income and they are often involved in forest-based income-earning activities.* Though women's cash earning may be substantial, it is not at the expense of the women's traditional responsibilities, which often means taking work home. The processing of many household products can be performed near home, thus allowing women to combine income-earning activities with household chores, such as childcare and food preparation. Women living below subsistence combine forest product gathering activities for their household use as well as for sale. Gathering and processing of food products may be favoured by women because of ease of access to forest resources; possibility of combining subsistence gathering with income-earning activities; flexibility of location of enterprise (near home) and their knowledge about forest products through subsistence use. The time spent gathering fuelwood leaves less time for cash-earning activities, while more expensively purchased fuelwood leaves less money for buying food. Furthermore, many income-earning activities, such as *fish-smoking* and *beer-making, depend on fuel wood.* As the price of fuelwood increases (either in labour or cash), the profitability of enterprise diminishes. Studies indicate that the constraints placed on women's income earning potential by fuel scarcity may have the *most harmful impact on food security for the household.*

FOOD AND NUTRITION SECURITY

Both during and after an emergency, access to adequate food, health and care may be severely constrained. *Household's nutritional status will be influenced not only by their access to an appropriate quantity and quality of food, but also by their access to wood or other fuel, clean water and food preparation equipment as well as time for feeding infants and young children.* Nutritional knowledge and cultural practices may influence the amount and the type of food that each person in the household receives. Lack of access to health care and sanitation as well as food also affects appetite, nutrient needs and the ability to absorb nutrients. The nutritional knowledge of household members is one of the factors influencing how they allocate their scare food resources among themselves and may determine whether *labouring adults* or *nutritionally vulnerable infants* and *children have priority* in receiving food. The ability to prepare suitable meals with the foods available is affected by households' understanding of their food and nutritional needs.

Food intake and nutritional status are inextricably interlinked. The impressive gains in global and national food supplies over the past 30 years were achieved because of effective investments in agricultural research. Although food security has improved globally, progress has been uneven. Significant proportions of the population are still food insecure in parts of Sub-Saharan Africa and South Asia. The adolescent female is a stakeholder that has not been a typical target of nutrition intervention. Thus we have fewer models of how to deliver preventive nutrition services to teenage girls. There is a critical need to identify and test interventions that effectively reach female adolescents, including the appropriate role of nutrient-dense dietary supplements.

FOOD SECURITY AND NUTRITION MONITORING

Food security and nutrition monitoring is undertaken to provide timely and relevant information to enable policy makers in formulating prudent policies and in implementing effective interventions to improve food security and nutrition. Food security and nutrition monitoring for the purpose of development planning and policy interventions addresses information needs of planners, policy analysts, and policy decision-makers at the local, regional and national levels. While information is generated largely by 1 or 2 sectors like agriculture and health, it is used by national planning agencies and policy analysis units.

Even with adequate capacity to implement food security and nutrition monitoring systems and to undertake food and nutrition policy analysis, no significant improvement could be made in the food security and nutrition situation unless issues relating the political economy are recognized and addressed to on a country-specific basis. It is also important to recognize the power structure and its influence towards policy reforms to make any headway in implementing food and nutrition policies.

HOUSEHOLD FOOD SECURITY AND NUTRITION IN AGRICULTURAL RELIEF AND REHABILITATION PROGRAMMES

Natural and human-caused emergencies can lead to rapid deterioration in food security and high rates of malnutrition. Relief aid cannot restore an area affected by famine to its previous situation or prevent future food insecurity; thus the need for early efforts to link relief, rehabilitation and sustainable development is compelling. Agricultural relief and rehabilitation activities should ensure access to food by vulnerable households and consider the nutritional needs of individuals.

During and after a crisis, food assistance may be a major source of nourishment. Nutritionists assess whether such aid is needed and ensure that the composition of the food provided is an appropriate one. Nutrition surveillance identifies individuals suffering from malnutrition and provides information for monitoring the impact of the emergency and the intervention. *Nutrition education* can enable families to *use unfamiliar foods* and influence the amount and the type of food that each person receives.

Information about local farming system and socio-economic conditions is used to address the causes of household food insecurity. The International Conference on Nutrition (ICN), convened by FAO and the World Health Organization (WHO) in Rome in 1992, identified strategies and actions for improving nutritional well-being and food consumption throughout the world.

Great strides have been made since the fifties in dealing with problems of poverty, hunger and malnutrition. Globally, food production has increased, the rates of pre-school malnutrition have fallen and infant and child mortality have decreased in most parts of the world. Despite these impressive gains, poverty, hunger and malnutrition continue to be problems of staggering proportions.

Nutrition interventions are not limited to food provision and consumption matters; a wide range of activities can have an impact on *food and nutrition security.* Provision of vegetable seeds can encourage even households with limited access to land to cultivate mixed gardens around their homestead. Home gardens can make valuable contributions to the *family in food supply* and *income* as well as providing a welcome complement to relief rations. Home-gardening is especially beneficial when attention is given to raising small animals as well.

POLICIES AIMED AT IMPROVING GROWTH AND DEVELOPMENT OF INFANTS

Breastfeeding is the optimal form of infant feeding. Health procedures need to provide an environment that actively encourages breastfeeding. Baby-friendly policies such as grooming in and early, continued contact between mother and newborn need to be actively pursued. This active promotion of breastfeeding needs to continue once the mother returns home. Breastfeeding support systems are a critical element for the successful continuation of breastfeeding. Where this support is not provided by the extended family, women recruited and trained from within the community can fill this void. An increasing problem in many countries is the entry of women into the formal wage-earning sector as a result of urbanization. While there are clear benefits for women from income-generation, continued breastfeeding becomes more of a challenge. Specific policies at the work site can facilitate the ability of women to breastfeed - breastfeeding rooms and flexible work schedules. These are not policies that typically have been high priority for many employers. Governments and the public health community need to collectively advocate for changes in the work environment that encourage breastfeeding.

POLICIES AND PROGRAMMES DIRECTED AT YOUNG CHILDREN

Most of the nutrition interventions that have traditionally been used by developing countries have targeted children. However, those interventions have done less well in reaching preschoolers under 3 years of age. Some promising approaches have emerged. Community-sponsored child care programmes that simultaneously address a mother's time constraint while providing a mix of health or nutrition services offer one model for reaching the younger child. Nutrition screening and monitoring services can be part of the package included.

Socio-economic conditions and cultural traditions have restrained women from acquiring managerial abilities to meet the needs of changing technology for socio-economic development.

COPING MECHANISMS ADOPTED BY RURAL WOMEN DURING ACUTE FOOD SHORTAGE

The food preparation, consumption and distribution-based coping mechanisms adopted by the rural women for food security during acute food shortage period like droughts include the following.

- Substitution of millets to rice and horsegram (*kulthi*) or cowpea to redgram
- Cooking only once a day
- Curtailment in number of meals from 3 to 2 by adults
- Selling the grains stored for seed purpose
- Reducing the quantity of food intake
- Curtailment in consumption of coffee and tea and diluting the milk or giving, coffee or tea to children instead of milk
- Preparation of *chutneys* in place of vegetable or dual preparations
- Use of more puffed rice
- Curtailing work and activities to suppress hunger
- And forceful starvation

❑❑❑❑❑

FOOD AND NUTRITION NEWS

Acharya N.G. Ranga Agricultural University

Vol. 3 *April 1998* *No. 1*

Mushrooms – For many uses

Mushroom is a simple form of fungus. It lacks chlorophyll and cannot therefore make its own food. It grows on dead organic matter or parasitically or symbiotically with other organisms. This constitutes an excellent means of recycling many of the farm wastes. Modern mushroom cultivation produces more proteins per unit area of land than by any other form of agriculture. Since these are grown indoors, no additional land is required for the cultivation. There fore mushroom cultivation is labour intensive and can provide employment to both educated and illiterate persons. They can be produced in large quantity with in a short time. The efficiency with which mushrooms convert carbohydrates into proteins is about 65% in comparison to about 20% for pork, 15% for poultry and 4% for beef.

More than 2000 species of fungi are reported to be edible throughout the world. In India about 200 of these are available. Eight of these have been cultivated so far. In all, about two dozen mushrooms have been cultivated for edible purposes in different countries. *Agaricus bisporus* (European or white button) *Volvariella sp* (Paddy straw or Chinese), *Pleurotus sp* (Oyster or tropical) and *Lentinus edodes* (Shiitake) mushrooms are the most popular ones among the commercial growers.

Of these, the first three are popular in India as techniques for their cultivation have been developed. In addition, *Agaricus bitorquis* (Edulis) *pholiotanemeko, Tricholoma matsutake, stropharia rugosoannulate, Temella fusiformis, coprinus fimetarius, Auricular*

polytricha, Agrocybe egerita and *Tuber melanospermum* are also being cultivated in different countries.

The total production of mushrooms in India is 11,600 tonnes, Uttar Pradesh, Himachal Pradesh, Haryana, Tamil Nadu and Punjab are the leading states and Rajasthan, Bihar and West Bengal are least producing states.

There are a number of mushrooms which are poisonous. It is often very difficult to identify the edible mushrooms from the poisonous ones. There are no prescribed standards to identify them. Some tests are often suggested, but none are reliable. It is stated that silver spoon or silver coin would turn blackish when they are dipped in a dish of poisonous mushrooms, exudations of milky substances from damaged fruit bodies, unusual changes in colour of onion while cooking with mushrooms etc., are some tests mentioned. Some of the highly poisonous ones are -*Amanita phalloides* (death cap), A. *pantherina* (Panther cap), *A. muscaria* (fly agaric), *Inocybe pattouirapdi, Collinarius speciosissimus, C. orellanus* etc.

Cultivation

Oyster mushrooms grow easily at 20-25° C and take a lesser time to reach maturity than others. Oyster mushrooms can be cultivated for the family table, since the harvest is quite bountiful, it can also prove to be a profitable hobby.

Different Stages of Oyster Mushroom Cultivation

Things needed include a tub, a few polythene bags (14 inches by 24 inches for home cultivation) and about 3 kgs of paddy straw. The seeds, called the spawn, are available at any nursery or with horticulturist, or pathologist.

Paddy straw is cut into two to three inches and soaked in tap water for 10 hours. Soaked paddy straw bits are drained well and boiled in fresh water for 45 minutes (sterilization). After sterilization, the paddy straw bits are drained of excess water and dried in shade. The straw bits are arranged in layers of 3 or 4 inches at the bottom of polythene bag. Two table spoons of spawn is sprinkled all along the periphery,' like a border. This process is operated till the bag is filled and the mouth of the bag is tightly tied which is known as 'bed'.

Two or three holes are made at the bottom of the bag for aeration. The beds are kept in cool and shady place (never in direct sunlight) for 15-20 days. During this period there will be mycelium growth. After 20th day polythene covers are removed. Mycelium growth has fused all layers of straw together into one compact cake-like block. A little water is sprinkled on it twice a day. At about 3rd or 4th day after opening the bag, oyster-like mushrooms sprout all around the bed. Fully grown mushrooms are harvested after 3 days. One bed will yield 4 to 5 crops.

Paddy Husk as a Substrate for Cultivation of *Plurotus Sajor-Caju*

Different substrates have been tried for the cultivation of several species of *Plurotus* in India and abroad. Mushrooms cultivated on two different substrates such as paddy straw and paddy husk in the Department of Foods & Nutrition, College of Home Science, Bapatla. The results indicate that the yield obtained from paddy straw was high (500g/bed) as compared with paddy husk (400g/bed). Paddy straw is considered as the best substrate for cultivating *plurotus sajor-caju* in terms of yield. As the cost of the paddy straw is increasing day by day it is advantageous to find out a low cost and alternate substrate for mushroom cultivation. **Vijaya Khader**

Economics

Economics of mushroom cultivation mainly depends on the equipments, nutrients used, their initial cost and yield performance. The expenditure for preparing the mushroom bed is Rs.10-15 which include the cost of the straw, polythene bags, expenditure towards labour charges, spawn and miscellaneous. One bed gives 1-2 kg of mushrooms from four pickings within a duration of 4 weeks. The market price varies between Rs.20-40/kg from place to place. The net income varies depending on the sale price of the spawn. If the sale price is Rs. 30/kg of mushroom the net income per bed after deductions is Rs.40-70. In the rural areas a women can easily manage 4-10 beds depending on the space available, helping them to earn Rs.180 to 450 per month as a supplementary income.

Nutritive Value of Mushrooms

Mushrooms are considered as delicacy as well as a good supplementary item of food. From the nutritional point of view, they are in between meat and vegetables

(Table I & 2). They provide high quality proteins and are low in calories. Hence, it is recommended as a special item for therapeutic use.

The protein in mushroom is rich in all the essential amino acids required by an adult. Some of the amino acids like tryptophan and lysine, which are absent in vegetable proteins are present in mushroom protein. Approximately 25-35 percent of total amino acid occur as free amino acids. They are also known to be excellent sources of riboflavin, niacin and pantothenic acid. The crude fat in mushrooms consistute less than one per cent. Mushrooms also contain appreciable amount of thiamine, folic acid and ascorbic acid. They are rich in minerals, and the mineral content in fresh mushrooms is higher than that in many fresh vegetables and fruits. Phosphorus and potassium are the main constituents, where as copper and iron are also present in appreciable amounts.

Table1

Proximate content of edible mushrooms - (Percentage fresh wt. basis)

Species	Moisture	Ash	Protein	Fat	Crude fibre
Agaricus bisporus	89.5	1.25	3.9	0.19	1.09
Lepiota species	90.0	1.09	3.3	0.18	0.86
Pleurotus sp.	90.0	0.97	2.8	0.65	1.08
Termitomyces sp.	91.3	0.81	4.1	0.22	1.13
Volvariella volvacea	88.4	1.46	5.0	0.74	1.38
V.diplasia	90.4	1.10	3.9	0.25	1.67

Table 2

Vitamin and mineral content of edible mushrooms (mg/l00g dry wt.)

Species	Thia-mine	Ribof-lavin	Nia-cin	Ascorbic acid	Ca	P	Fe	Na	K
Agaricus bisporus	1.1	5.0	55.7	81.9	23	1429	0.2	–	4762
Lentinus edodes	7.8	4.9	54.9	0.0	33	1348	15.2	837	3793
Pleurotus ostreatus	4.8	4.7	108.7	0.0	98	476	8.5	61	–
Volvriella volvacea	1.2	3.3	91.9	20.2	71	677	17.1	374	3455

Nutritive Value of Oyster Mushroom at Three Different Stages of Maturity

Mushrooms were collected from two different places and analysed the nutrients like protein, fat, crude fibre, ash, calcium, iron, phosphorus and energy for the different stages of both white and grey varieties of oyster mushroom as per the standard methods of A.O.A.C. Carbohydrate content was calculated.

White and Grey varieties of oyster mushrooms were used in the present study. Three different stages of maturity such as immature (two days before harvesting), mature (at the time of harvesting) and dried (matured and sun-dried) stages were used.

Significant difference was observed between white and grey varieties of mushrooms with respect to their nutrient content, except crude fibre. No significant difference was observed between immature and mature stages of both varieties. Dried mushroom contain high nutrient contents when compared with immature and mature stages because during drying process the food also losses moisture and there is an increase in concentration of nutrients in the remaining mass. **A Padmavathi and Vijaya Khader**

Bio-Availability

Studies on *invivo* digestibility of mushroom protein carried out on young male albino rats at 10% level of protein revealed 87% of over all digestibility of mushroom protein as against 93% for edible soya bean cake, which is significantly different at 5% level. Apparent digestibility of mushroom protein is very low (37%) as compared to edible soya bean cake (84%) which may be due to the use of raw dried mushroom powder being used in rat diets. **Vijaya Khader**

Effect of cooking on nutritive value

Moisture and vitamin C content decreased on maturation of *Volvariella volvacea* mushrooms (paddy straw). Mature buttons were cooked by three common methods namely (a) boiling in water for 10 min (b) shallow frying in refined oil and (c) deep frying in refined oil for 10 minutes. Among the three methods of cooking, the loss of nutrients were more in shallow frying, followed by deep frying and minimum in boiling. Vitamin C losses varied from 50-75% based on the type of cooking. Heat treatment did not affect the crude fibre, fat and ash contents. **P. Shobha Devi, G. Sarojini**

Food Uses

Therapeutic Uses

There are many therapeutic effects attributed to mushrooms. It is true that only a fraction of this has, been so far exploited. Many of the folklores in different countries provide ample evidence for the therapeutic use of mushrooms from very ancient times.

Hypolipidemic Action

Inclusion of paddy straw mushroom powder in the hypercho-lesterolaemic diets, (64% starch; 10% ground nut oil; 15% protein; 4% salt mixture; 1% yeast powder; 5% cellulose and 1% cholesterol) reduced the plasma total lipids, total cholesterol and glyceride level significantly whereas free fatty acids and phospholipid levels were not much affected. The lowering of total cholesterol (TC) level was equally reflected both in HDL and LDL + VLDL fractions. Hypolipidemic action of *Volvariella volvacea* could be due to the increased catabolism or decreased absorption as evident by the lower plasma levels without lipid accumulation in liver inspite of higher cholesterol intake.

Heart lipids (total lipid, cholesterol, free fatty acids, phospholipids and glycerides) decreased with inclusion of mushroom powder. But effect was significant only in case of glycerides and phospholipids. Moreover, the lipid lowering was more evident only at 5% level of mushroom powder in the diet than that at 10% level.

Vijaya Khader

Hypoglycemic Action

The effect of fibre from oyster mushroom, spinach and isabgol on blood glucose response was studied in twenty four non-insulin dependent diabetic patients (NIDDM) from a diabetic clinic. The experimental diets included a normal breakfast, *pulkha*, mushroom powder incorporated *pulkha*, spinach powder incorporated *pulkha* and isabgol incorporated *pulkha*. Two levels of fibre i.e., 10% and 15% were studied. The breakfast was served with cucumber curry and a cup of coffee made with skim milk powder without sugar for a period of three days. Blood samples were drawn before and after fibre rich breakfast consumption and analysed for plasma glucose. Mean peak rise over fasting was significantly lower when compared with the test and normal (low fibre) breakfast. Of the two fibre levels tested 15% fibre from mushroom was found to be effective.

J. Alakananda and Vijaya Khader

Weaning Foods

Mushroom powder can be effectively incorporated in weaning foods which serves as useful protein source and also contributes major nutrients like, protein, and calcium.

The chemical analysis indicate the weaning food I is better as compared with other two weaning foods in respect of protein, fat, calcium and calorific value. All the three weaning foods are quite acceptable without any remark. The keeping quality of weaning foods was quite satisfactory up to a period of five months, as indicated by the acceptability studies. The microbiological observations showed no drastic changes in the weaning foods at the end of fifth month.

Preparation of Weaning Foods with Mushrooms

Weaning Foods

Weaning food I

Puffed rice powder	– 20g
Whole Bengal gram (germinated, dried roasted).	– 20g
Mushroom powder	– 20g
Jaggery	– 10g
Cardamom	– One

Preparation : Above ingredients were mixed thoroughly and packed in a clean dry glass bottle with cover.

Weaning food II

(*a*)	Quantity	(*b*)
Rice powder	– 40g –	Kodra powder
Red gram dal powder	– 10g –	Green gram dal powder.
Bengal gram dal powder	– 10g –	Bengal gram dal powder
Mushroom powder	– 40g –	Mushroom powder

Preparation : The cereal or millet, pulse and mushroom powders were mixed and prepared batter and fermented overnight. Jaggery 10g, salt to taste and pinch of soda were added and mixed well and baked at 150°C for 10-15 minutes. Allowed to cool at room temperature dried well and powdered, packed in a clean dry glass bottle with cover.

Vijaya Khader and Nayana Pandya

Culinary Uses

With the idea of popularising the mushrooms among the local communities, some recipes have been prepared and evaluated for their acceptability. The overall acceptability of chutney powder was excellent compared to pickle and ketchup. The keeping quality of the preparation were good upto a period of 2 months. Subtle in flavour, characteristic of biting, texture and attractive in appearance, mushrooms lend themselves to several dishes, some of which are : *Mushroom soup, Mushroom curry; creamy mushroom, Mushroom pizza, Mushroom stuffed capsicum, Mushroom pan cakes, Mushroom pakoda, Mushroom omlette, Mushroom sandwiches, Mushroom Ketchup, Mushroom chutney, Mushroom pickle, Baked mushrooms, Mushroom flavoured rice, Mushroom garlic sauce, Mushroom flakes, Mushroom chips* and *Mushroom instant soup powder.* They also go well with *noodles, macaroni* and vegetables like cauliflower, peas, potatoes and tomatoes.

On account of being cereal based, Indian diet suffers from inadequate protein content. Mushroom protein has been universally accepted to be superior to vegetable proteins and as good as animal proteins. Mushrooms are therefore, recognised as non-conventional source of protein which can bridge the protein gap in the Indian diet. Mushrooms are essential supplements to the cereal based diet and go a long way in making it a healthy diet. As, it is rich in fibre, low calorific value and absence of cholesterol, mushrooms find application in diet therapy (treating obesity, diabetes and various heart diseases).

Pickle

The most common method of preserving mushroom is preparation of pickle. Two types of pickles can be prepared using fresh mushrooms. The details of preparation are as follows:

Mushroom Pickles

Mushrooms are fried in oil and all the spice powders were mixed well.

Pickles were kept at room temperature for a period of 5 months and regular observations were carried out to see the shelf life. Acceptability studies indicate that the scores for taste, flavour and overall acceptability for both the pickles are markedly different from each other. The second pickle stored well for 5 months without fungal growth. The first pickle got spoiled with in a week. The microbiological observation of spoiled pickle showed the presence of *streptococci* and *staphylococci.*

Pickle I		Pickle II	
Mushroom	125g	Mushrooms	125g
Cumin seed powder	1/2 tsp	Mustard seed powder	1/2 tsp
Fenugreek seed powder	1/2 tsp	Turmeric powder	1/2 tsp
Coriander seed powder	1/2 tsp	Chilli powder	1/2 tsp
Turmeric powder	1/2 tsp	Salt	1/2 tsp
Mustard seed powder	1/2 tsp	Asafoetida	Pinch
Green chillies	5Nos	Sesame oil	30 ml
Vinegar (Commercial)	45 ml		
Sesame oil	45 ml		
Salt	to taste		

BOOKS PUBLISHED BY Dr. (MRS.) VIJAYAKHADER

Sl. No.	Name of the book	Address
1.	Money from Mushrooms (English)	Information & Communication Centre A.N.G.R.A.U. Rajendranagar Hyderabad 500 030.
2.	Mushrooms (Telugu)	Information & Communication Centre A.N.G.R.A.U. Rajendranagar Hyderabad 500 030.
3.	Mushrooms for Livelihood (English)	Kalyani Publishers, 1/1, Rajinder Nagar, Ludhiyana-141008.
4.	Puttagodugulu (Telugu)	Sri Lakshmi Press, Bapatla-522 101. Guntur District.

❑❑❑❑❑

EFFECT OF FIBER FROM MUSHROOMS, SPINACH AND ISABGOL ON BLOOD GLUCOSE RESPONSE IN NON-INSULIN DEPENDENT DIABETICS

The cardiothoracic Journal Vol. 3, No. 5, July, 1997, PP-31-40

Miss Jwala Alakananda and Dr. Vijaya Khader
Post Graduate and Research Centre, Rajendra Nagar, Hyderabad-30

Abstract

Diabetes mcllitus-a metabolic disorder- is the third commonest disease in the world next to cardiovascular an oncological disorders, which is being treated with drugs and more recently with dietary modifications. The effect of fiber from oyestcr mushrooms, spinach and isabgol on blood glucose response was studied in twenty-four non-insulin dependent diabetic patients. Blood samples before and after fiber rich breakfast consumption were analysed for plasma glucose tolerance test in comparison with normal (low fiber) breakfast. Of the two fiber levels tested, 1 5 g fiber from mushrooms was effective, whereas, even 10 g fiber from spinach and isabgol was sufficient in controlling postprandial blood glucose levels effectively.

Introduction

Diabetes mellitus — a heterogeneous, hugely genetically determined disorder— is associated with relative or absolute de- fluency of insulin. In India, the prevalence rale is 2-4% of which 90% are non-insulin dependent diabetics. The fundamental abnormality in diabetic metabolism is impaired utilization of glucose and one of the major aims of diabetes therapy is to normalize for every patient the blood glucose profile. Though blood glucose levels can be lowered with the help of drugs, such an approach is never free from side effects. The most suitable approach would be the one involving changes in dietary regimes.

According to diabetes control and complications trial (DCCT) "tight control'" of glycaemic status of an individual is associated with reduction of diabetic complications'. Treatment not only consists of administering pharmaceutical agents but also tailoring an individualized schedule of exercise, diet and lifestyle. This is in the true sense a "holistic approach" to the treatment for every patient.

Introduction into the diet of carbohydrates in a form that is absorbed slowly in the small intestine or not at all has been advocated as a simple means of preventing or treating diabetes mellitus Complex carbohydrates result in lower glucose and insulin responses than equivalent amount of simple carbohydrates.

Persons with diabetes mellitus may greatly benefit from increasing their intake of dietary fiber which refers to the endogenous components of plant material in the diet which are resistant to digestive enzymes produced by man. Fibrous material exerts direct physiological effects throughout the gastro-intestinal tract in addition to affecting metabolic activities indirectly. The interplay of these effects is responsible for the beneficial effect on the carbohydrate metabolism through a series of mechanisms, which are related to the type and physico-chemical nature of the fiber.

Oyester mushrooms [Pleurotus citrino pileatus] are a novel source of food picking up a strong base in the present market. Mushrooms, because of their therapeutic value and high digestibility arc gaining access to in the present day healthy diet. They have no cholesterol, are almost fat free and are excellent sources of vitamins and minerals. Major portion of carbohydrates in them is in the polymeric form constituting the dietary fiber. Richness of mushrooms in dietary fiber is nutritionally recommeneded for consumption to mitigate the chances of hyperglycaemia.

Spinach (Spinacea oleracea) is a low cost, nutritious commonly consumed and easily procurable green leafy vegetable which is known for its fiber content, apart

Table 1. Total dietary fiber content in dehydrated oyester mushrooms and spinach

Source of fiber	Dietary fiber content (%) (g/100 g)		
	TDF	IDF	SDF
Oyester mushrooms	39.9	39.15	0.75
Spinach	42.99	31.74	11.25
*Wheat flour	1.9	NA	NA

NA : Not Available

Note : TDF - Total dietary fiber

IDF - Insoluble dietary fiber (*Nutritive Value of Indian Foods, ICMR, 1993)

Table 2. Neutral detergent fiber content in dehydrated oyester mushrooms and spinach

Source of fiber	Neutral detergent fiber content (%) (g/100 g)
Oyester mushrooms	45.5
Spinach	33.0

from vitamins and mineral content. It is rich in calcium, iron, B-carotene, vitamin-C, ri-boflavin and folic acid. Spinach contains a good amount of fiber mucilage and is considered to be helpful in controlling blood glucose levels in diabetics. Since the whole vegetables, and not their purified fiber, are the rich sources of vitamins and minerals, it is advised to include these in large amounts, rather than their purified fiber.

No studies are available on the hypoglycncmic effects of mushrooms and spinach on human beings. Hence, the present study was undertaken to find out the possibility of using them as hypoglycaemic agents.

Materials and Methods

Dehydrated oyester mushrooms and spinach samples were estimated for neutral detergent fiber content and total dietary fiber content. Based on the results of fiber estimation, the required amount of raw material (oyester mushrooms and spinach) was calculated.

Study period. The total study period was 21 days.

Selection of subjects. Twenty-four non- insulin dependent diabetics were selected from a center for diabetes

All subjects were in the age range of 40-55 years. Moderate diabetics i.e., patients with diabetic history of 3-7 years on oral drugs only were selected. They were healthy and free from any other major complication.

Experimental diets. The experimental diets included a normal breakfast (Pulkha), mushroom powder incorporated pulkha, spinach powder incorporated pulkha and isabgol incorporated pulkha. Each of these was served with cucumber curry and a cup of coffee made of skimmed milk and without sugar.

Experimental groups. Subjects were divided into two groups of 12 each. The first group of subjects (group A) received breakfast providing fiber from the tested sources at a level of 15 g and the second group of subjects received breakfast items providing fiber from the same sources at a level of 10 g. The breakfast items were planned in such a way so as to provide 1/3rd of the daily calorie requirements. Since, all the subjects belonged to the same age range and taking average physical exercise their caloric requirements remained almost similar and were computed based on ICMR recommendations.

Determination of glyenemic response. The subjects were requested to report at the pathology laboratory after an overnight fast and also asked to avoid taking drugs 24

hours prior to the test. Fasting and post-prandial blood samples were drawn intravenously after consumption of glucose and test items and plasma glucose levels were estimated using "Monozyme Glucose Method". Glycaemic response of each breakfast was calculated taking area under the two-and-half hour glucose stimulation curve (AUC) as index value,

$$\text{Glycaemic response} = \frac{\text{AUC of breakfast item}}{\text{AUC of glucose}}$$

Table 3. Patterns and schedule followed for providing various fiber incorporated breakfasts

Fiber source	Level at which It Is incorporated (g)		Duration of feeding the breakfast
	Group A	Group B	
Mushrooms fiber	15 (37.5)	10 (25)	3 days
	Followed by 3 days normal diet		
Spinach fiber	15 (34.8)	10 (23.25)	3 days
	Followed by 3 days normal diet		
Isabgol fiber	15 (15)	10 (10)	3 days

Note. Values in parenthesis indicate the amount of respective fiber source that would contribute to 15g and 10g of fiber respectively.

Table 4. Mean plasma glucose values of the subjects to the different breakfast items (15g and 10g fiber)

Items tested	Plasma glucose (mg%)					
	Fasting	hr	1 hr	1 hr	2hr	2 hr
Glucose 15 g	115.91	165.16	195.83	202.75	183.41	115.83
(n=12)	±15.06	± 17.18	± 18.59	± 101.59	± 23.91	±24.67
l0g	114.91	183.66	212.58	250.00	186.08	116.33
(n=12)	±15.37	±54.55	±20.18	± 32.33	± 36.51	±40.00
Normal B15g	118.08	179.16	207.41	173.33	140.41	122.33
(n=12)	±14.92	±17.28	±21.37	±15.26	± 11.78	±12.89
l0g	122.91	192.33	221.25	184.66	149.58	125.83
(n=12)	±15.20	±21.10	±29.34	24.19	±12.22	±13.94
M.I.B. 15 g	115.08	149.16	177.58	157.58	148.75	116.75
(n=12)	±16.90	±11.50	±21.04	± 21.78	± 20.75	±12.80

10g	118.16	158.33	189.16	172.58	162.50	120.50
(n=12)	±17.89	±23.40	±34.34	±34.30	±31.05	±15.68
S.I.B. 15g	120.41	157.25	194.25	170.58	157.41	125.58
(n=12)	±25.08	±12.30	±24.88	±21.41	±19.20	18.23
10 g	116.66	161.91	198.41	179.58	166.75	123.33
(n-12)	±20.30	±24.60	±37.69	±36.28	±27.32	±17.88
I.I.B. 15 g	116.00	143.00	165.75	146.91	137.83	115.91
(n-12)	±15.51	±17.84	±24.13	±22.43	±41.91	±15.71
10 g 118.00	145.91	173.25	153.066	142.16	120.08	
(n=12)	±25.19	±29.93	±34.67	±35.83	±33.43	±23.31

Note.	MID - Mushrooms incroporated breakfast SIB - Spinach incropornted breakfast IIB - Isabgol incroporated breakfast

Results, discussion and conclusion

Results showed that the mean peak rise over the fasting blood glucose values after consumption of the incorporated breakfasts is less when compared to glucose both at 15 g and 10 g levels. At both levels, the mean peak rise is least with isabgol incorporated breakfast, followed by mushrooms incorporated breakfast, and spinach incorporated breakfast when compared with the respective mean peak rise with glucose values. The highest values of blood glucose levels among all the breakfast items are noticed with normal (low fiber breakfast) at both 15g and 10g fiber levels but are lower when compared to glucose.

Of all the breakfast items tested, the glycaemic response is lowest with isabgol followed by mushrooms incorporated breakfast and spinach incorporated breakfast. The glycaemic response of normal breakfast (low fiber) is found to be the highest of all food items tested and almost similar to that of glucose.

At half hour after stimulation, no significant difference is observed between glucose and normal breakfast and also within fiber enriched breakfasts; but significant difference is observed with mushrooms and isabgol incorporated breakfast when compared to glucose and normal breakfast. One hour alter stimulation, there is no significant difference between glucose and normal breakfast and also glucose and spinach, but a significant difference is observed with isabgol and mushrooms when compared to glucose and also between spinach and normal breakfast.

At one and half hours alter stimulation, a significant difference is observed between glucose and various breakfasts and there is no significant difference between mushrooms and isabgol and also between spinach and mushrooms. At 2 hours after stimulation, although there is no significant difference between the various breakfasts, a significant difference is observed when the breakfast items are compared to glucose, and the trend is similar at 2 1/2 hours after stimulation.

The best control of plasma blood gulcosc levels on an average is with isabgol incorporated breakfast followed by mushrooms incorporated and finally spinach incorporated breakfast. On an average, when peak values over fasting are observed, no significant difference is observed when the breakfast items are compared to glucose. A significant difference is observed when mushrooms and isabgol incorporated breakfasts are compared with normal (low fiber) breakfast.

Table 5. Mean glycaemic response of food items

Food item	Glycaemic response
Glucose	1 .00
Normal breakfast	0.92 + 0.07
Mushrooms incorporated breakfast	0.84 + 0.11
Spinach incorporated breakfast	0.86 + 0.11
Isabgol incorporated breakfast	0 77 +0.14

Table 6. Effect of fiber from breakfast items on plasma glucose response at different time intervals

Time interval	Glucose	N.B.	M.I.B.	S.I.B.	I.I.1J.
1/2 hour	174.4bc	153.7a	153.7a	159.5ab	144.4a
1hr	204.2cd	214.3d	183.3ab	196.3bc	169.5a
1/2 hr	226.3a	179.0ab	165.0ab	175.0ab	150.2a
2hr	184.7c	145.0ab	155.6ab	162.0b	140.0a
Peak rise	122.3d	93.8C	66.7ab	77.7bc	52.5a

N.B. : Normal breakfast
M.I.B. : Mushrooms incorporated breakfast
S.I.B. : Spinach incorporated breakfast
I.I.B. : Isabgol incorporated breakfast

High fiber foods would result in improved glucose control due to delayed digestibility and absorption and also by increasing mouth to caecum transit due to the presence of higher proportions of celluloses and hemicelluloses". Leafy vegetables contain more of uronic acids and fiber present in them is of insoluble nature (celluloses and hcmicclloloses). The beneficial effect of fibre in diabetics could be attributed to its effect on increased efficacy of endogenous insulin to metabolize glucose. Also when glucose is absorbed at the distal end of the small intestine, the release of gastric inhibitory polypeptide is in small amount which acts to potentiate the insulin responses to a carbohydrate load, thus causing less strain on B cells of pancreas and simultaneously regulating blood glucose values.

Thus, it can be concluded that dietary liber plays an important role in controlling the blood glucose levels in diabetic patients and prevents hyperglycacmic peaks. Mushrooms which arc a rich source of liber exhibited significant results, similar to spinach, liven though isabgol was much more effective it is always advisable to include the natural sources of fiber from foods and not from supplements. Significant differences were observed between 15g and 10g fiber in case of mushroom incorporated breakfast, and 15G liber is more effective in controlling the blood glucose levels; wheieas. in case of spinach and isabgol no significant differences were observed between the two fiber levels, suggesting that even 10g fiber from the testcd sources is sufficient lor effective blood glucose control. Therefore, oyester mushrooms and spinach can be recommended for diabetics to be included in their diets for effective Mood glucose control.

Summary

Addition of plant fibers to diets of patients With diabetes is accompanied by significant reduction in post-prandial hypcrglyeacima. In the present study. The effect of fiber from mushrooms, spinach and isabgol was studied in twenty-lour non-insulin dependent diabetic patients. Results obtained indicated that the mean peak rise over lasting plasma glucose levels was significantly lower alter all the tested items, when compared to oral glucose tolerance test. Isbagol and spinach incorporated breakfasts showed effective blood glucose control at 10g fiber level, whereas in case of mushrooms 15 g fiber elicited better blood glucose control. Therefore, mushrooms and spinach can be recommended for diabetics to be included into their routine diets lor effective blood glucose control.

❑❑❑❑❑

HYPOLIPIDEMIC ACTION OF VOLVARIELLA VOLVACEA IN ALBINO RATS

Vijaya Khader Indian Journal of Clinical Practice Vol. 4. No. 3. August 1993 PP 42 - 48

Inclusion of sun dried Volvariella volvacea (Paddy straw mushroom) at 5% or 10% level in hypercholesterolemic diet resulted in insignificant effect on gain in body weight, inspite of higher food intake. This also resulted in increase in food efficiency ratio of diets. Feeding of Volvariella volvacea also caused a number of lipid changes in plasma, liver and heart. Lipid, cholesterol and glyceride levels in plasma were significantly decreased without any effect on free fatty acids and phospholipids. Hepatic lipid profile was not much affected by feeding Volvariella volvacea whereas total lipid levels of heart significantly decreased. Cholesterol, phospholipids and glycerides were also reduced,but effect was significant only in the case of phospholipids and glycerides.

Introduction

Hypercholesterolemia/hyper-lipidemia is one of the suspected pathogenetic factors in the development of atherosclerosis - coronary heart disease connected with thickness of arteries 1, 2. Though blood cholesterol level can be lowered with the help of drugs either by inhibiting the endogenous synthesis or by lowering cholesterol absorption from the intestines, such an approach is never free from side effects. The most suitable approach would be the one involving changes in dietary regimens. The ability of various edible fungi to lower blood cholesterol was first reported by Kaneda et al[3] and Kaneda and Tokuda[4]. Lowering plasma lipids by the consumption of certain foods including mushrooms which also offer other attractions have, therefore, a certain appeal. Since there is very limited information on the hypochoteslerolemic and hypolipidemic effects of mushrooms, the present study was undertaken to find out the possibility of using Volvariella volvacea as a hypocholesterolemic and hypolipidemic agent.

Materials and Methods

Rat experiments: Male albino rats, 5-6 weeks old, were divided into 3 groups (A,B and C) and were housed individually. The rats of group A (control group) were fed on hypercholesterolemic diet containing 64% starch, 10% groundnut oil, 15% casein protein, 4% salt mixture, 1 % yeast powder, 5% cellulose and 1 % cholesterol. Animals

Table I: Gain in body weight(g) and food intake of rats fed on hypercholesterolemic diets containing dried mushroom (means ± SEM)

Diet groups*	Initial body weight(g)	Final body weight(g)	Gaininbody wieigh(g)	Food intake	FER
A Control	78.14±4.20	104.14±5.23	26.00±1.30	215.60±6.41[a]	0.121±0.004[a]
B	76.33±4.53	99.72±7.22	23.39±1.56	365.30±18.50[a]	0.064±0.003[a]
C	74.83±5.55	98.33±6.53	23.50±2.82	398.33±3.33[a]	0.059±0.002

A- Hypercholesterolemic diet (basal Diet + 1 % Cholesterol)

B- Hypercholesterolemic diet + 5% dried mushroom powder

C- Hypercholesterolemic diet + 10% dried mushroom powder

a- Significantly different from control (P<0.01)

Furthermore, inclusion of dried mushroom powder in the diet had no significant effect on organ weight index (Table - II)

Table II: Organs weight of rats fed on hypercholesterolemic diets containing mushroom powder (means ± SEM) Weight g/100 g body weight

Diet groups*	Lung	Heart	Spleen	Kidney	Liver
A (Control)	0.937±0.083	0.466±0.031	0.375±0.034	0.896±0.065	3.678±0.315
B	1.058±0.083	0.577±0.150	0.322±0.035	0.896±0.049	4.535±0.084
C	1.037±0.134	0.443±0.018	0.346±0.041	0.847±0.063	4.160±0.198

* *as in Table-I*

Table III: Levels of Lipids (mg/dl) in plasma of rats fed on hypercholesterolemic diets containing dried mushroom powder (means ± SEM)

Diet groups*	Total lipids	Cholesterol	Free Fatty	Phospholipids acids	Glycerides
A control*	515.2±2.51	158.9±13.4	37.53±2.60	84.61±2.08	234.0±7.62[b]
B	429.8±25.2[a*]	112.±8.2[b**]	33.39±1.90	76.4±1.20	204.6±5.70
C	481.5±9.79	151.0±10.0	37.90±2.70	84.5±3.70	212.20±8.90

*as in Table - I * P<0.01 ** P<0.05*

Table IV: Cholesterol levels (mg/100ml) in plasma lipoprotein in rats fed on hypercholesterolemic diets containing dried mushroom powder (mean ± SEM)

Diet groups*	TC	HDL	(LDL+VLDL)C	HDLC:TC
A(Controi)	158.9 ± 13.40	79.9 ± 2.24	79.0 ± 3.41	0.487 ± 0.035
B	112.0 ± 8.24^{b}	57.5 ± 1.61^{a}	54.4+2.76^{b}	0.513 ± 0.029
C	151.0 ± 10.0	66.2 ± 1.85^{a}	84.8 ± 4.45	0.438 ± 0.022

* *as in Table-I a: P<0.01 b: P<0.05*

Table V: Liver Profile (mg/g) or rats fed on hypercholesterolemic diets containing dried mushroom powder (mean ± SEM)

Diet groups*	Total lipids	Cholesterol	Free Fatty acids	Phospholipids	Glycerides
A Control)	48.29±2.42	16.89±2.20	17.77±2.77	11.90±0.69	1.73±0.52
B	46.49±2.21	17.63±1.14	15.68±2.03	8.60±1.10	4.59±1.84
C	46.00±1.61	17.96±2.40	17.54±1.84	8.36±0.59	2.13±0.60

* *as in Table I*

of group B and C were given hypercholesterolemic diet containing 5% or 10% dried powdered Volvariella volvacea, respectively. Records of diet intake and gain in body weight were maintained weekly. Food Efficiency Ratio (FER) was calculated as follows.

$$\text{FER} = \frac{\text{Gain in body weigtht}}{\text{Food intake}}$$

After 4 weeks of feeding, the rats were sacrificed and their blood was drawn by direct cardiac puncture. Plasma was separated by centrifugation (3000 × g) for 10min and stored in refrigerator till further use. The tissues like lung, liver, kidney, spleen and heart were removed and washed immediately with normal saline to remove extraneous matter. After blotting, fresh weights of tissues were recorded and these were stored in a deep freezer until further analysis.

Biochemical analysis: Lipid changes including total lipids, total cholesterol[4], high density lipoprotein (HDL), cholesterol Phospholipids[6], free fatty acids[7] were determined in plasma and tissues. Gfyceride levels were calculated by subtracting phospholipid, cholesterol and free fatty acid values from total lipid values. Cholesterol value of low density lipoprotein + very low density lipoprotein (LDL+VLDL) was calculated by subtracting HDL - cholesterol values from total cholesterol values.

Results

It is evident from the present study (Table I) that inclusion of Volvariella Volvacea in hyper cholesterolemic diet at either of the levels did not show any effect on gain in body weight. The food intake was, however significantly (P<0.01) higher in mushroom fed groups resulting in a decreased food efficiency ratio.

Inclusion of mushroom powder in the hypercholesterolemic diets reduced the plasma total lipids, total cholesterol and glyceride level significantly whereas free fatty acids and phospholipid levels were not much affected (Table -III). More over, the lowering of total cholesterol (TC) level was equally reflected both in HDL and LDL + VLDL fractions. The ratio of LDL cholesterol to total cholesterol was not significantly effected on mushroom feeding (Table -IV)

There was no significant effect on mushroom inclusion in the diets on the hepatic total lipids, free fatty acids, cholesterol and phospholipids as compared to control (Table-V). However, glyceride levels were slightly higher but the effect was insignificant.

Table No.VI: Heart lipid profile (mg/g) of rats fed on hypercholesterolemic diets containing dried mushroom (mean ± SEM)

Diet groups*	Table lipids acids	Cholesterol	Free Fatty	Phospholipids	Glycerides
A (control)	49.50±1.21	2.41±0.24	15.64±1.41	5.50±0.55	24.90 ± 1.24
B	35.35±1.05[a]	1.90±0.09	11.74±1.14	3.77±0.12[b]	18.24±1.35[b]
C	40.00±2.45[b]	2.15±0.06	17.50±0.30	3.55±0.25[b]	16.71±0.86[b]

* as *in Table -I a: P<0.01 b: P0.05*

The results indicate that feeding of dry mushroom powder causes no effect on the accumulation or mobilization of lipids in the hepatic tissue when fed with hypercholesterolemic diets.

Heart lipids (total lipid, cholesterol, free fatty acids, phospholipids and glycerides) decreased with inclusion of mushroom powder (Table-VI). But effect was significant only in case of glycerides and phospholipids. Moreover, the lipid lowering was more evident only at 5% level of mushroom powder in the diet than that at 10% level.

Discussion

The findings indicate that the food intake was significantly higher in mushroom fed groups, which resulted in lower values of food efficiency ratio. Since there was no effect on the organ weight indices of rats fed on hypercholesterolemic diets containing sun dried mushroom powder revealed no drastic effects even when fed continuously for 4 weeks at high level.

Inclusion of Volvariella Volvacea In the diet had a significant lipid/ cholesterol lowering effect. Miche et al(1970) and Tokila et al (1972) have also shown the hypocholesterolemic effect of Shiitake, a Japanese mushroom. That it might be due to the presence of lentysine in Shiitake. Kaneda and Tokuda (1966) have reported that dried mushroom preparation markedly reduces plasma cholesterol levels and all caps and stems were effective to various degrees of lowering.

Since there was no lipid or cholesterol accumulation in the liver it indicates that hypocholesterolemic/hypolipidemic action of Volvariella Volvacea could be due to the increased, catabolism or decreased absorption as evident by the lower plasma levels without lipid accumulation in liver inspite of higher cholesterol intake.

References

1. Ames, B.N. (1966) Assay of inorgonic phosphate, total phosphate and phosphates. Methods in Enzymology. Academic Press, New York.
2. Bruzell, J.D., Schnott, H.G., Motielsky, A.G. & Bierman, E.L,- (1976) Myocardial infaction in the familial forms of hypertriglyceridemla. Metabolism. 25, 313-320.
3. Hulley, S:E., Rosenman, R.H., Bawal, R.D., & Brand, R.J., (1980) Epidemiology as a guide to clinical decisions; the association between triglycerlde and coronary heart disease. New England Journal of Medicine, 302,1382-1389.
4. Kaneda, T., Arai, K., and Tokuda, S., (1964). The effect of dried mushrooms, cortinollus shiitake on cholesterol metabolism in rats. Journal of Japanese, Society of Food and Nutrition, 16, 106-108.
5. Kaneda, T., and Tokuda.S., (1966). Effect of various mushrooms preparations on cholesterol levels in rats. Journal of Nutrition, 90, 371-376.
6. Lopes - Virilla, M.F. Stone, P., Ellis, S., and Colwell, J.A., (1977). Clinical Chemistry, 23, 882 cited from Micro analysis in Medical Biochemistry (ed.F.D.P. Watton & H.Freeman).
7. Lowry, R.R., & Tinsley, I.J., (1976). Rapid Calorimetric determinations of free fatty acids. Journal of Amcerican Oil Chemical Society, 53, 470.
8. Michi, K., Sakurai, S., & Kurinane, W., (1970). Isolation of a hypocholesterolemic substance from Shiitake mushroom Lentinus edodus. Journal of Japanese Society of Food and Nutrition, 23,218-222.
9. Zak, B., Zidrkia, A., & Boyce, A.J. (1953). A new methods for the direct determination of serum cholesterol. Journal of Laboratory and Clinical Medicine, 41, 435-437.

❑❑❑❑❑

MUSHROOM CULTIVATION OFTERS LUCRATIVE BUSINESS

G. Venkataramam

The Hindu, Thursday, Feb. 11, 1993

There has been a virtual boom in mushroom production, consumption and export in Madras in the last two years.

Mushroom cultivation is an ideal avocation of the landless, unemployed people, and it assures a return of over two hundred per cent in just 45 days. In no other venture can one get this attractive rate of returns,said Mr. K. Kalyanasundaram and Mr. S. Dharmaraj, the young entrepreneurs, who deal with oyster mushroom cultivation and marketing in Madras.

"We started a small farm and training centre to popularise mushroom cultivation and marketing about five months ago, and we have so far trained about 55 people. Mostly unemployed youth, housewives, small-scale entrepreneurs, retired armed forces personnel came to attend the training programmes, and major industrialists approached us enquiring about the commercial viability of the project and the market potential," said Mr. Kalyanasundaram.

After a brief training-cum-demonstration, the participants, who enrol themselves by paying Rs. 150, are given two bottles of oyster mushroom (Pleurotus sp.) spawn, five polythene bags and the other necessary ingredients to grow mushroom in their own homes. "We do a systematic follow-up, and see to that each one of them grows at least one kg mushroom a bag. We buy fresh produce at Rs. 30 a kg. Thus the participants get their money invested on training back from the first crops in less than 50 days," said Mr. Dharmaraj. The subsequent crops add to their profits.

With the market assured and the technology so simple more and more people are entering in to this agri-business. A large number of training institutes have sprung up, and they also offer good package of technologies and services and ensure a ready market at the site of production. "The technology has been so simplified that even lay people can grow mushrooms with great success. Then again there is a growing market offering attractive returns. The services offered and the follow-up work done by the entrepreneurs have made the oyster mushroom cultivation click with the people, explained Mr.Kalyanasundaram and Mr. Dharmaraj. Their farm markets about 600 kg fresh mushroom each month.

Export prospects

There are more than 30 such institutes and units in the city, which play a positive role in popularising the mushroom cultivation. On an average about 500 kg fresh mushroom is produced each day, and about a tonne of the processed mushroom comes from here. Fresh mushrooms sell for Rs. 40 to 45 a kg in the market, and dried ones fetch up to Rs. 500 a kg in the market. About 10 kg fresh mushrooms will make a kg of dried one. The dried mushrooms are preferred for exporting and the bulk of the 'mushrooms is exported to Singapore, Malaysia, France, Germany, the Eastern European countries and West Asia.

Among the various kinds of mushroom, the oyster mushroom has been found ideal for growing in the plains and the coastal regions It thrives best in 20 to 25 degrees Celsius, can be maintained with relative ease in the tracts. Only in summer months, the growers will have to take special care in lowering the temperature and controlling the humidity. Unlike the other mushrooms like the button mushroom, which can be grown in the cool hilly climate, the robust and delicious oyster mushroom has a wider adaptability, and can be grown with less care.

The economics of cultivation

"The economics of mushroom cultivation is quite inviting. The cost of cultivation per kg mushroom works out to Rs. 10 for 45 days, and the value of the produce is Rs. 30 at the growing the site. I started cultivating mushrooms after attending a training course in November, and started with 18 beds. I harvested 9 kg fresh mushrooms and sold in the ready market at Rs. 30 a kg. Now I am confident of expanding the scale of operation, and want to enter into dried mushroom trade with an investment of Rs. 30,000," said Mr. P. Ravindran, a young and enthusiastic entrepreneur of Royapettah. The techiques are so simple and the support services and market price are encouraging, accordfng to Mr. Ravindran.

There are also some constraints in the mushroom business. There is generally a drop in production during summer, and it leads to a wide fluctuation in the demand and price. Though, there has been a growing consumer preference for the mushrooms, there has been some reservation about its consumption in certain quarters. Again, not many are aware of the various mushroom recipes. Presently it is used mainly in soups, certain forms of curry and in fried rice. But this protein-rich fungal vegetable can be used just like cauliflower in different delicious dishes.

"Through sustained awareness generation campaign the nutritional properties of mushrooms should be popularised in order to make it acceptable to a wider cross-section of the society," said Mr. Kalyanasundaram. Mushrooms are a rich source of high quality vegetable protein, and being a low calorie diet it is recommended for diabetics and other health conscious people. Besides, mushrooms have a host of medicinal properties.

The existing marketing structure needs to be fine-tuned in order to keep the price level steady. An organised marketing network will be a boon to the growers as well as the consumers, and it will also help in streamlining the exports. There is need to step up mushroom processing (drying), as it will enhance the storage life of the mushrooms and fetch rewarding returns. Fresh ones keep well for only two days, and this perishable nature forces the growers into distress selling at times of glut.

Special efforts should be made to link up all small growers, and prevent drop-outs during summer. This would ensure a regular supply to the market and stabilise prices. The training and services extended to the growers should be standardised and stringent quality control measures introduced to make mushroom cultivation an organised commercial enterprise for the unemployed and housewives in urban areas.

EDIBLE OYSTER MUSHROOMS ON COTTON PLANT STALKS

Prof. R.H. Balasubramanya (2006)

Indian Farming, PP 26–27

Oyster mushrooms can be artificially cultivated on various lignocellulosic crop residues, viz. rice straw, wheat straw, cotton stalks etc. This mushroom can also be grown on composted materials. The yield varies from substrate to substrate and in any case it is around 500-600 g/kg of raw material without any nutrient supplementation. The yield can be increased with supplementation.

Mushrooms are the fleshy sporophores of fungi known to row in nature on decaying cellulosic materials, dead wood, soil and manure its. Majority of these fungi belong to the class Basidiomytina and a few to the class Ascomycotina. Edible fungi under the order Agaricales and the amilies Agaricaceae, Polyporaceae and luteaceae have been under commercial cultivation.

The edible mushrooms are delicacy a food and form one of the choicest able dishes. They are rich in protein and an excellent source of vitamins and minerals. Most of the mushrooms have a very low starch content and can form an ideal food for diabetic patients.

Though, about 2,000 species have been deported to be edible, the techniques of artificial cultivation are available only or a dozen of species of which about four are being cultivated on a commercial scale. These are, *Agaricus bisporus* (white button mushroom), *Volvariella* spp (tropical mushroom or paddy straw mushroom), *Lentinns edodes* (Japanese mushroom) and *Pleurotus* spp (Oyster mushroom). The button mushroom requires low temperature and grows on fermented substrates, whereas paddy straw mushroom grows on unfermented substrates and at an elevated temperature of 35°C. The Japanese and oyster mushrooms also grow on unfermented materials at temperatures of around 20°C and 30°C, respectively. *Pleurotus* spp have the ability to grow on both fermented and unfermented materials.

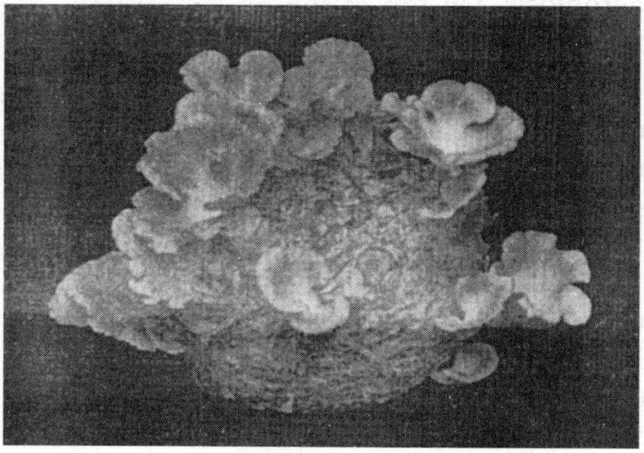

Mushroom can be artificially grown on cotton plant stalks

Being a tropical country, India has been bestowed with bountiful crops suiting to all three different seasons. Crop residues are available throughout the year. The diversified mushroom cultures available in the country can be best adopted to different places as per the temperature conditions and raw material availability. For successful growing of oyster mushrooms, the substrate has to be processed, i.e. cut to about 3.0-4.0 cm size and pasteurized at around 80°C to ward off the competing moulds. This is an energy intensive step. Secondly, the spawn preparation using mostly the grain spawn is very common. The quality of the spawn decides the yield of sporophores. The yield of fleshy fruiting bodies is normally around 500g/kg of the substrate. The yield of *Pleurotus* on different substrates is given in Table 1.

Pleurotus spp can be successfully cultivated at temperatures of around 30°C and hence is becoming increasingly popular in both tropical and sub-tropical countries. In India, the cultivation of this mushroom is picking up at an alarmingly high rate due to ease of its cultivation. The most important step in the cultivation of this mushroom is the pasteurisation of huge quantities of raw material without which the competing moulds during spawn run could not be eliminated. Presently, the procedure adopted is to cut the raw materials to about 3-4 cm, soak them in cold water for 10-12 hr, followed by hot water treatment (about 80°C) for 2-4 hr. In addition, the practice of adding fungicides bavistin and formaldehyde to the substrate is also followed. It has been shown that the anaerobic treatment at room temperature replaces the hot water treatment. Using an anaerobic treatment plant constructed, studies were undertaken for one full year and data collected established that hot water replacement is completely effective.

ANAEROBIC TREATMENT

A batch type anaerobic digester (1 m dia and 0.6 m depth) was constructed underground with brick masonary construction. Water seal was provided at the top of the digester in which an MS lid (18 SWg, 1.25 mm thickness) was placed to create anaerobiosis. On top of the lid, an outlet was provided to release the gas produced during anaerobic treatment.

The cut materials of rice straw and cotton stalks were charged in the digester containing mixed microbial consortium as inoculum. Care was taken to see that the entire material was under water. This was achieved by placing a perforated lid. Thereafter the digester was covered by placing the lid in the water seal and the hose pipe taken out from the outlet was dipped in water. In the case of rice straw the treatment was for 24 hr, whereas it was for 48 hr in the case of cotton stalks. The anaerobically treated materials were given a wash in freshwater containing alkali (0.1% based on the original weight of the raw material) and the excess water was drained out by placing on a perforated wire mesh. The materials thereafter were packed in polythene bags by sprinkling grain spawn in each layer by the conventional polythene bag technique. The bags were incubated for spawn run and subsequently the mycelial entangled bags were hung for the production of mushroom sporophores.

HOT WATER TREATMENT

To compare the efficacy of this new method, the conventional hot water treated materials were also used for seeding the mushrooms.

The interesting feature of this inexpensive anaerobic treatment is that the resident microflora which normally cannot be dislodged from the surface of the cellulosic materials by soaking overnight in water are removed with the digester fluid. The organisms, if any, adhering loosely to the surface at the end of the treatment are removed by a simple dilute alkaline water wash. Thus, the number of organisms are reduced to a significant extent. The usual molds encountered as competitors, viz. the species of *Trichoderma* and *Penicillium* during the conventional method were not found to grow on the anaerobically treated materials.

The yield of fleshy mushroom fruiting bodies of P. *sajor-caju* is about 500-600 g on an aerobically pre-treated rice straw and cotton stalks which is akin to the hot water processed materials. Blending of rice straw with cotton stalks (75:25) reduces the spawn run period from 20 days to 15 days, whereas the period is only 10 days when only cotton stalks are used.

CONCLUSION

The anaerobic treatment being ecofriendly and devoid of hot water run, farmers can readily adopt this technology on the farm itself to process the raw materials for growing oyster mushrooms. It is possible to process 100 kg of the raw material every day by installing two digesters. Depending on the retention time, the material can be fed every day or alternate day. The materials taken out are to be washed in freshwater, drained and to be spawned. The requirement of 1,000 litres of hot water at 80°C to pasteurise 100 kg material can be dispensed with. The initial investment incurred on the construction of semi-continuous plant or batch fed plant could be recovered in less than one year, since the energy to heat the water to 80°C is totally avoided. In view of this, the current method becomes inexpensive and ecofriendly as no fungicides are used.

Table 1. Yield of *Pleurotus* on different substrates*

	Mushroom species			
Substrates	**P.sajor-caju**	**P.florida**	**P.flabellatus**	**P.ostreatus**
Wheat straw	400	410	417	440
Rice straw	310	425	307	453
Pearl millet straw	365	440	362	513
Cotton stalks	480	453	450	600

*g/kg of the substrate

❑❑❑❑❑